Popular Deceptions

What they haven't told us and how much it's going to cost

Acknowledgements

This book wouldn't be possible without the encouragement and patience of my wife, family, and friends. Untold thanks to all of you! My thanks to Mr. John Grimwade who wisely advised that we forget any preconceived ideologies and biases, keep digging until we understand the fundamental facts, and then let those facts guide our thinking.

I would also like to acknowledge the Kansas City Writers Meetup Group for their inputs and support. This is a wonderful group of authors willing to freely share their knowledge, experience, and encouragement whether you're a first-time author or a well-established writer.

My thanks to Hazel Hart, author of the young adult novel, *The Survivalist's Daughter*, and three suspense novels. Hazel persevered through countless proofreads and a barrage of endless questions to help bring this book to fruition. Finally, a special thanks to Gordon Kessler, the author of fifteen books, including *Novel Writing Made Simple*, the thrillers *Brainstorm*, *Dead Reckoning*, and *Jezebel*, and "The E Z Knight Reports" series. Gordon served as editor, formatter, and coach through the final process of publication.

Thank you all for your kind, knowledgeable support and wisdom.

Table of Contents

Part I
The Elements of Deception

"All I know is just what I read in the papers,
and that's an alibi for my ignorance."
—Will Rogers (1879-1935)

Chapter 1
Facts Are Facts

"You can fool all the people some of the time, and some of the
people all of the time,
but you cannot fool all of the people all of the time."
—Abraham Lincoln

We humans love our facts. Facts tell us about the world around us. They tell us how things work and why things work the way they do. By definition, facts are true, which is one of the many reasons we love our facts. It's also what makes facts a powerful means of persuasion—and deception.

Want to win the next casual debate among friends? If so, all you have to do is offer a few official-sounding statistics and say, "Facts are facts." You can then walk away knowing you've won the debate. You know you've won because, after all, no one can argue with the facts. Or can they?

This won't shock anyone, but facts have a problem: not all facts are created equal. They come in varying degrees of truthfulness and completeness. Some facts simply don't tell us enough to be of any real value. Some facts are more misleading than informative. Some are only true under a specific set of circumstances. Others simply don't tell the whole story.

As an overly simplistic example, if someone tells us that grass is green, we know it's true because we've all seen green grass. At the same time, however, we know this true fact is also false much of the time. When grass goes dormant during winters in northern climates, it turns brown. Knowing that grass is green doesn't tell us the whole story, and it's only true under specific circumstances.

Consider one of the most infamous types of facts in existence— statistics. Former British Prime Minister Benjamin Disraeli said,

"There are three kinds of lies: lies, damned lies, and statistics." A more colorful description compares statistics to bikinis; "What they reveal is intriguing, but what they hide is critical."

Statistics are without question the most often misused, misapplied, and misinterpreted types of facts we encounter. As an example, between the mid-1940s and the early 1970s, statistics showed that average global temperatures steadily declined. Based on this *evidence*, every major climate organization on the planet was warning us to prepare for the coming ice age.[1]

We also understand that statistics can be easily biased and manipulated by cherry-picking the data included or excluded in the calculations. That much is obvious from watching commercials for pickup trucks—every brand of truck can't provide the highest horsepower, greatest towing capacity, and best gas mileage.

Statistics are also frequently used to *prove* that a cause-and-effect relationship exists between two separate events. This so-called proof is often so insignificant, it borders on the ridiculous. The alleged connection may be little more than pure coincidence. For example, statistics *prove* that the more firemen called to a fire, the greater the damage. Does that mean firemen are causing the damage, or does it simply mean we send more firemen to the most damaging fires? Statistics also show that when ice cream consumption increases, the number of people who drown also increases. Does that mean eating ice cream increases your chance of drowning, or does it simply mean we eat more ice cream in the summer when more people swim?

This next statistic will raise more than a few eyebrows. From 1980 through the end of 2008, the more coal we burned to generate

---

1. Steven Goddard, "1970s Global Cooling Scare," *Real Science*, http://stevengoddard.wordpress.com/1970s-ice-age-scare/ (Accessed 12-5-13).

electricity in the US, the cleaner our air became.[2] While this statistic is absolutely true, it's doubtful anyone is going to believe that burning more coal will clean the air. Suffice it to say we understand the dangers of accepting statistics as proof of anything. Despite this understanding, statistics continue to exert a powerful influence on our beliefs.

There's another type of fact that is even more troubling and more deceptive than statistics—the silent facts, a.k.a. *missing* facts. Intentionally excluding facts that don't promote your product or support your point of view is the first rule of marketing. It's a rule that's lavishly applied. Whether watching a product commercial, a public service announcement, or the evening news, it's often the missing facts that are the most critical.

Picture Deceptions

If a picture is worth a thousand words, graphs can be powerfully persuasive because they provide a picture presentation of the facts. Like statistics, however, graphs are easily manipulated to hide the truth, exaggerate trends, or simply fool us all. For example, let's assume Figures 1-1 and 1-2 on the next page show the population of a rural town in the year 2000 and 2010. Both graphs plot the exact same two numbers, but they yield decidedly different perceptions.

---

2. US Environmental Protection Agency (EPA), "National Emissions Inventory (NEI) Air Pollutant Emissions Trends Data," http://www.epa.gov/ttnchie1/trends/ (Accessed 10-23-13); and Energy Policy Institute Data Center, "Providing a Plan to Save Civilization," Coal Consumption in the United States, 1950-2012, with Projections for 2013, http://www.earth-policy.org/index.php?/data_center/xls/indicator12_2010_6.xls (Accessed 12-9-13).

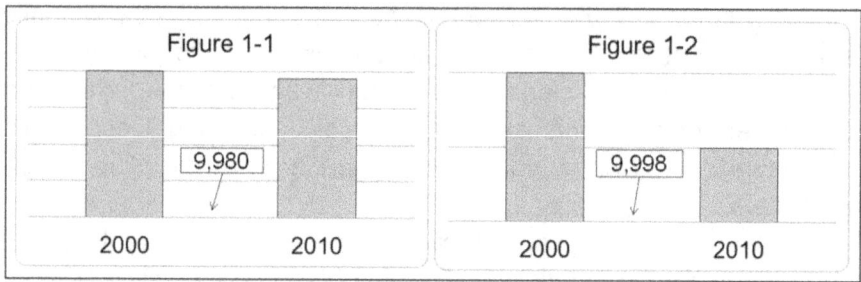

In Figure 1-1, it appears the population declined by a very small margin, but in Figure 1-2, it appears the town lost 50% of its residents. The actual change was a drop from 10,000 residents to 9,999. The town lost one resident. How are these two graphs possible if they plot the exact same two numbers? All it takes is the creative selection of values used for the graph's vertical scale.

In this case, both vertical scales plot a high value of 10,000, but neither Figure uses zero at the bottom of the scale. The lowest value plotted in the first graph is 9,980 and it's 9,998 in the second graph. With 9,998 as the lowest number graphed, the loss of one resident represents a 50% reduction—obviously a gross exaggeration of reality.

Figure 1-3 on the next page is not meant to be a political statement about the environment. It's shown because it's an example you may have seen and because it demonstrates how easily graphs can exaggerate a trend.

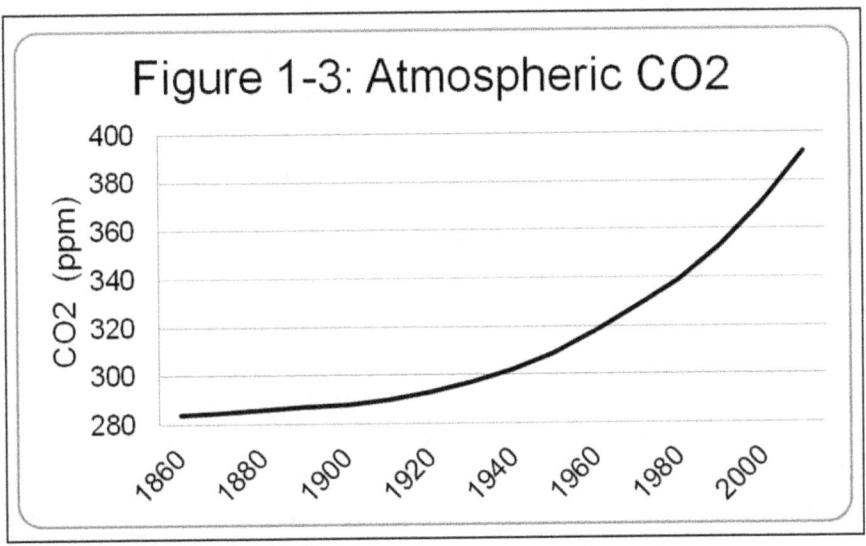

Figure 1-3: Atmospheric CO2

This is the most often used graph I've found regarding increasing levels of carbon dioxide (CO_2). Like the earlier figures, the vertical scale on this graph doesn't start at zero. For the casual viewer, the increase in CO_2 appears to be 10,000%, but in reality it's closer to 40%. That's more than enough to raise our concern, so the graph's exaggeration is unnecessary, but effective. Certainly, graphs can be invaluable tools for identifying trends or explaining complex data, but they can also be extremely misleading.

Ordinary Facts

We can refer to facts like the ones discussed above as *ordinary facts*, *half-truths* or *missing* facts. In the chapters that follow, we'll see real-world examples of these types of facts being used to deceive; we'll see how such facts have cost consumers billions of dollars; and how, in more than a few cases, they've led to unnecessary human suffering and even death.

The Fundamentals

Fortunately, there are facts that aren't as easily manipulated or as easily biased as the facts discussed above. There are facts that are always true, facts that do tell the whole story, and facts that actually explain cause-and-effect relationships. We can refer to these facts as *fundamental facts* or simply *the fundamentals*.

Returning to our overtly simple example of grass, there's one fundamental fact that explains everything we need to know about the color of grass: chlorophyll. During the growing season, grass uses chlorophyll for photosynthesis. Chlorophyll is green, which gives grass its green pigmentation. When grass is dormant in winter, the chlorophyll moves to the roots for storage, taking the green pigmentation with it.

Once we understand the fundamentals behind any issue, we'll find that we no longer lose many arguments—at least not with rational people. When we understand the fundamentals of an issue, we'll also find that it's easy to know when *ordinary* facts don't tell the whole story. Fortunately, as demonstrated by the green grass example, we don't need to be rocket scientists to understand fundamental facts. The fundamentals can be simple—and they're enormously powerful!

Batteries and Fundamental Facts

We can understand the power of the fundamentals by considering a popular commercial used to promote car batteries. This commercial takes advantage of two well-known problems: car engines can be hard to start in cold weather, and car batteries lose power in extreme cold. Imagine being the last one to leave work on a snowy, subzero night and realizing your battery is too weak to start your car. It's a situation no one relishes.

Taking advantage of these issues, the commercial in question shows a battery frozen in a solid block of ice, yet the battery starts a car with no difficulty. Amazing, right? Not so much, at least not if we remember the fundamental facts we learned long ago. At an early age, we learned that water freezes at thirty-two degrees Fahrenheit. This battery isn't very cold. In many parts of the world, temperatures hovering around thirty degrees represent a balmy winter evening. We should also note that the car in this commercial was sitting under bright lights in a Hollywood studio, so its engine wasn't cold.

The *ordinary* facts in this commercial tell an impressive story— a solidly frozen battery still has enough power to start a car. The *fundamental* facts tell a very different story—there's nothing special about a near-thirty degree battery being able to start a warm car engine. When we take time to consider the fundamentals, we're not impressed, but you can bet this commercial sold a lot of car batteries.

Beyond Commercials

Twisting the facts isn't limited to product advertisers, politicians, and Big Business. Our trusted public interest groups are also guilty. As an example, consider promotions for renewable electricity from technologies like wind turbines and solar panels. It's likely you've heard claims indicating that renewable electricity is a key step to freeing ourselves from foreign oil.

It's an effective promotion for a number of reasons, but it fails to consider one fundamental fact provided by the US Department of Energy (DOE):

> Only 1% of the oil consumed in the US is used in the production of electricity.[3]

The reality is that oil is *not* an energy issue: It's a transportation and chemical industry issue. Electricity and oil (transportation) are two completely separate concerns requiring different public policies to adequately address each one. Unfortunately, that's not how our thinking or our policies have evolved.

If we installed all the renewable electric energy we could ever want or need, we'd only reduce oil consumption 1%. This is a fundamental reality repeatedly ignored by the news media, environmentalists, and politicians when touting the virtues of renewable electricity.

This isn't a knock against renewable energy. It's a knock against half-truths. In this instance and many others, so-called experts are telling us to expect a result they know can't be achieved. Is this an accidental oversight or is it intentional deception?

Obviously, this issue won't stop our support of renewable electricity. That's not the point. The point is we're being deceived all

---

3. US Department of Energy (DOE), Energy Information Administration (EIA) "2008 Annual Energy Review,"
http://www.eia.doe.gov/emeu/aer/pecss_diagram.html (Accessed 4-16-11).

the time—even when we think we understand the issue. This simple deception is symbolic of the half-truths that shape our beliefs regarding a number of our most critical public policies.

Consider our perception of the logging industry. It's among the most reviled industries in existence due to its impact on the environment, yet it provides the raw materials needed to support jobs for millions. In the US, sawmills alone have annual payrolls of nearly $3 billion. Roughly eleven million private landowners rely on the lumber industry to provide or supplement their incomes.[4] While the economic role of this industry is significant, that's not why it's mentioned here. It's mentioned because the fundamental facts regarding US logging tell a very different story than the one we've grown to believe.

According to studies submitted to the United Nations (UN) Food and Agriculture Organization, forest growth in the US has exceeded lumber harvests since the 1940s; in 1997, the volume of forest growth was 380% greater than in 1920; and the standing volume of wood per acre is some 33% greater today than in 1952.[5] In addition, 82% more hardwood trees are growing today than forty years ago.[6]

This is tremendous news. Loggers in the US are not only maintaining sustainable forests, they're increasing the size and the quality of our forests and they've been doing so for some seventy years! Of course, this good news is based on those dreaded statistics, which begs an important question. How do we know when we can or can't trust the information we're given? Sadly, there's no sure-fire way to know—unless we do our homework and learn the fundamentals.

For example, those wonderful-sounding statistics about US

---

4. US Lumber Coalition, "About the US Lumber Industry," http://www.uslumbercoalition.org/general.cfm?page=31 (Accessed 1-9-13).
5. The North American Forest Commission Twentieth Session; St. Andrews, New Brunswick, Canada (June 12-16, 2000), "State of Forestry in the United States of America," http://www.fao.org/docrep/meeting/x4995e.htm (Accessed 8-21-12).
6. USDA Forest Service, "Forest Resources of the United States," General Technical Report (1993), American Forest & Paper Association (1992), http://www.associatedpallet.com/timber/forestfacts.htm (Accessed 11-12-11).

forests may be less impressive than they appear. Pioneers clear-cut essentially every tree they could access during the westward expansion of the US. Given this starting point, it wouldn't take a great deal of effort to show tremendous improvements. No one is suggesting we all go hug a lumberjack. Deforestation remains a major worldwide problem, and the US contributes to that problem by importing lumber. Still, it's good to know that forest growth in the US has outpaced lumber harvests since the 1940s.

The point to this example is simply this: We seldom hear the whole story and what we do hear often hides the fundamental truth. The news media is far more likely to stir outrage over plans to cut down the city's oldest elm tree rather than mention the success of our efforts to protect forests.

About Facts

We're very glad someone is bringing attention to deforestation, we're not upset by creative promotions for car batteries, and we'll continue to support renewable electricity even if it has a minor impact on our need for oil. These examples are minor and of little consequence. However, when taken to the extreme, or when seemingly minor deceptions continue to accumulate regarding important issues, the deception is no longer minor.

In the chapters that follow, we'll see countless examples of deception. We'll see ordinary facts that hide simple truths. We'll reveal missing facts that refute many popular beliefs. More importantly, we'll see that the deceptions are often intentional—and they're impacting your wallet and mine.

Chapter 2
Science Sells

"If we knew what it was we were doing,
it would not be called research, would it?"
—Albert Einstein

You might win friendly debates using facts, but if you really want to clinch the victory, cite the results of an independent scientific study. We can argue the facts all day long, but as laypersons it's tough to argue against science. The marketing gurus of the world understand this. They know their message will be more persuasive if their claims are supported by independent science—even if that science is biased and wrong. Suffice it to say, "Science sells."

Like facts, however, science has a few problems of its own. Not all science is created equal, not all independent science is truly independent or unbiased, and not all scientific findings are correct.

Historically Bad Science

Did you know that scientists initially developed heroin to treat respiratory disease? In 1898, a major pharmaceutical company began the commercial production of heroin as a routine medical treatment for respiratory conditions, including the common cough.[7] The medical world soon realized that patients developed a tolerance for the drug, making it less effective after repeated treatments. The addictive power of heroin also became better understood as the new drug was tested. International treaties soon followed to ban the production and use of heroin.

This "medical breakthrough" turned out to be a scourge on

---

7. Hosztafi S., Acta Pharm Hung, "The History of Heroin" (August 2001), http://www.opioids.com/heroin/historyheroin.html (Accessed 3-4-10).

society that is still impacting the world today. Scientists promoting heroin as a cure for the common cough didn't fully understand the fundamental and tragic issues tied to heroin. The science was incomplete.

Medical practitioners in the nineteenth century also thought they knew how to cure childhood hyperactivity. Soothing syrups were marketed to help stressed-out parents calm their rowdy children. These syrups often contained morphine, chloroform, opium, or cannabis. Sometimes they included combinations of these *soothing* substances. [8] Scientists knew the syrups accomplished their intended purpose, but they didn't anticipate the potential negative consequences, such as addiction, lethargy, and the likelihood of children being harshly punished for staring incessantly at the *pretty flowers* (the weeds) rather than pulling those weeds out of the vegetable garden as instructed. The science was incomplete.

In the late 1800s, cocaine was considered a natural wonder drug.[9] Coca-Cola originally contained cocaine, and it wasn't until the early 1900s that Coca-Cola became completely cocaine free.[10] An effective marketing slogan might have read, "Need a late afternoon pick-me-up to last until quitting time? Drink Coca-Cola!" Today the best we can do to survive the afternoon grind is to drink one of those high-powered energy drinks laced with caffeine and energy-generating herbs. The scientific understanding of cocaine was incomplete.

Heroin for coughs, soothing syrups for rowdy children, and cocaine in soft drinks: Perhaps now we know what our grandparents and great-grandparents meant when they said they longed for the good old days!

---

8. Nathan Birch, "The 10 Most Insane Medical Practices in History" (Nov 20, 2007), http://www.cracked.com/article_15669_10-most-insane-medical-practices-in-history.html (Accessed 2-14-10).
9. Eva Pronovost, "When and Why Was Cocaine Removed from Coca-Cola?" *Culinary Arts 360* (July 31, 2008), http://www.helium.com/items/1132039-coca-cola-cocaine-when-was-cocaine-removed-from-coca-cola-did-coca-cola-contain (Accessed 5-31-10).
10. Barbara Mikkelson, "Cocaine-Cola," *Snopes.com* (May 19, 2011), http://www.snopes.com/cokelore/cocaine.asp (Accessed 2-14-10).

Lobotomies[11]

Do you remember lobotomies (no pun intended)? Conceived in 1935 by Dr. Egas Moniz, lobotomies were thought to be a cure for many mental disorders, including schizophrenia and such common behaviors as mood swings. At one time, doctors envisioned a world where lobotomies would become a simple office procedure, much like a trip to the dentist. Apply a little Novocain, place an ice pick under the eyelid and behind the eyeball, drive the ice pick into the brain with a rubber mallet, and wiggle it around. The wiggling action would disable selected parts of the brain that were malfunctioning. The patient could be home in time for dinner.

That's a rather crude recounting of the process, but you get the idea. It was a gruesome procedure and a barbaric approach to mental health, but it was backed by the best science of its time. It was considered a medical breakthrough. Dr. Moniz won the 1949 Nobel Prize in Medicine for his work in developing the lobotomy procedure.

The science was incomplete, but that didn't stop 20,000 lobotomies from being performed in the US in the 1940s and '50s. In Canada, lobotomies were performed as late as 1961. At times, the lobotomy procedure actually succeeded in correcting the desired mental disorder. Unfortunately, the process was often successful at correcting other *disorders*, such as the ability to speak, walk, or feed one's self. Many patients were left in a vegetative state. Dr. Moniz's work may have paved the way for advances in modern brain surgery, but at what cost? The science was incomplete.

Speaking of Brain Cells

Have you ever been told the adult brain doesn't grow new brain cells? It's not true, but it's what scientists believed until the late

---

11. Margarita Tartakovsky, "The Surprising History of Lobotomies." *PsychCentral, World of Psychology,* http://psychcentral.com/blog/archives/2011/03/21/the-surprising-history-of-the-lobotomy/ (Accessed 5-4-13).

1990s.[12] Today we know the adult brain continues to grow new brain cells. That's great news for those who enjoy an occasional adult beverage because we all know alcohol kills brain cells. Actually, that's not true either. Heavy alcohol use can damage the brain's ability to function, but abstinence restores lost function.[13] Today's latest science indicates that moderate alcohol consumption actually improves memory and cognitive skills (reasoning).

The popular belief that alcohol kills brain cells can be traced to the temperance movement, a powerful special interest and political lobby in the late 1800s and early 1900s. This political lobby promoted an anti-alcohol agenda. To promote their agenda, the temperance movement combined the latest scientific knowledge regarding alcohol's impact on the brain with influential images of stumbling skid row alcoholics. The message was highly influential and the movement grew so powerful that Congress passed the 18th Amendment outlawing the manufacture, transportation, and sale of alcohol. The Amendment was later repealed—once *sober* minds prevailed. The marketing campaigns against alcohol were so highly effective that today, some eighty years after the repeal of Prohibition, most of us still believe moderate alcohol consumption kills brain cells.

The temperance movement was a well-intentioned, nonprofit organization that had our best interests at heart. It grew to become a highly influential political powerhouse—it's not easy to pass a Constitutional Amendment. As we'll see in upcoming chapters, the temperance movement wasn't the last well-intentioned, nonprofit organization to gain such power and influence. It's also not the last one to use the latest science and influential images to shape public opinion.

---

12. Dr. Emily Senay. "Adults Can Grow Brain Cells," November 2, 1998, *CBSNEWS*,   http://www.cbsnews.com/news/adults-can-grow-brain-cells/ (Accessed 11-14-12).
13. Dr. Michael E. Charness, "Abstinence Restores the Alcoholic Brain," Harvard Medical School published in *Journal Watch*, Neurology (March 20, 2007), Taken from Bartsch AJ et al. "Manifestations of early brain recovery associated with abstinence from alcoholism." *Brain* 2007 Jan; 130:36-47, http://www.jwatch.org/jn200703200000001/2007/03/20/abstinence-restores-alcoholic-brain (Accessed December 19, 2009).

Today's Science

Today's science is light-years ahead of the error-filled science discussed above, but no less vulnerable to being wrong. By definition, science is observation and experimentation that leads to theories. If a theory agrees with all historic and current observations and if the theory withstands the test of time and the intense scrutiny of other scientists, the theory becomes the "generally accepted" explanation of why and how an observed event happened. The trouble is scientists never know when their observations are complete. Therefore, scientists can never be 100% sure they've got it right and we can never be certain the science is complete. True scientists understand this limitation and acknowledge that even well-supported scientific theories never serve as absolute proof.

Many would argue against this line of reasoning. After all, we know that scientists can mix two chemicals, and every time they do, they'll get the same chemical reaction. We're 100% certain we can prove that this chemical reaction will always occur, right? It's not that simple. Many chemical reactions change in zero gravity, and they simply don't occur at absolute zero—the lowest theoretically achievable temperature.

Gravity provides another example of scientific observations that fail to serve as final proof. How can that be? Newton nailed the science of gravity way back in 1687 with his Law of Gravity. This isn't just a theory. It's so well documented it's considered a Law of Nature. Newton's Law is a mathematic formula that accurately predicts the attraction between two objects with mass. The trouble is Newton's Law of Gravity doesn't work for objects traveling at or near the speed of light or objects that have very low mass (subatomic particles). Newton was absolutely correct based on everything within his view, but his observations were incomplete. His theory doesn't tell the whole story, and it's only true under specific circumstances. Today, we're exploring new particles and new theories, including quarks, gluons, and the Higgs boson to expand our understanding of gravity. Will we ever be certain that our observations are complete?

Our search for life on other planets also highlights the limitations of science. Scientists have always assumed that life on other planets would require water, a tolerable level of gravity, a temperature range similar to Earth, and six key elements including phosphorous—yes, phosphorous. The reasoning was simple: Every living thing we've ever observed required these things. Because science is based on observations, scientists had no other choice. They were forced to conclude that these things are required in order for life to exist. Our observations changed in December 2010.

That's when scientists announced they found new bacteria on Earth that was missing one of the key elements required for life: phosphorous. Instead of phosphorous in this bacteria's DNA, scientists found arsenic, a poisonous substance to most living things.[14] This was no minor scientific discovery. It opened new possibilities in our search for life on other planets. The science regarding the essential requirements for life was incomplete—or was it?

Later studies could not identify arsenic in this bacteria's DNA and concluded that phosphorus was still required for this life form to exist.[15] The amount of phosphorus required to sustain this bacteria, however, was described as shockingly low and the fact that this bacteria thrives in arsenic-rich environments is a departure from past observations. The point remains the same: scientists are limited by what they can observe and they can never be certain their observations are complete.

An Epidemic of Bad Science

An article in *Scientific American*, "An Epidemic of False Claims," by Professor John P.A. Ioannidis, describes an explosion of false and exaggerated scientific findings in today's peer-reviewed science. The article indicates that the number of researchers,

---

14. Richard A. Lovett, "NASA Life Discovery: New Bacteria Makes DNA With Arsenic" (December 2, 2010), *National Geographic Daily News*, http://news.nationalgeographic.com/news/2010/12/101202-nasa-announcement-arsenic-life-mono-lake-science-space/ (Accessed 5-3-13).
15. Richard A. Lovett, "Arsenic-Life Discovery Debunked—But 'Alien' Organism Still Odd," *National Geographic Daily News* (July 7, 2012), http://news.nationalgeographic.com/news/2012/07/120709-arsenic-space-nasa-science-felisa-wolfe-simon/ (Accessed 7-15-13).

studies, and experiments is increasing exponentially. As a result, the competition to publish is fierce and often leads to shortcuts in the research and exaggerations of the validity or the importance of the study's conclusions. Also as a result, the conclusions are often wrong and they're often more misleading than informative. The real concern is that the media is eager to publish the sensational findings of such research, but we seldom hear the limitations or the errors behind the findings.

Professor Ioannidis also points out that today's science is filled with false findings because a great deal of today's research is not being pursued to find the truth. Many so-called *independent* studies are far from independent. The research may be well-intentioned, but conflicts of interests and underlying biases often corrupt the study from start to finish[16]. Think of the temperance movement funding a study on the effects of alcohol on the human brain. How much faith should we have in the results of that study?

Two additional articles on the subject of bad science are worth mentioning: *Lies, Damned Lies, and Medical Science* by David H. Freedman and *How Facts Backfire* by Joe Keohane. The titles speak for themselves; the latest scientific research is often biased and wrong.

There's another reason we see so many false scientific claims. Many of today's scientific studies are little more than exercises in statistics. This is especially true for the abundance of medical findings that dot the headlines. It's exceedingly rare to find a study that actually discovers a direct link between a medical aliment and the cause of that aliment. Instead, the findings are more often based on statistical analysis of massive volumes of data. The statistical results may or may not prove anything at all; however, if the study's findings can create a sensational headline, it's likely the mainstream media will warn us about the newly *proven* threat to our health.

---

16. John P.A. Ioannidis, "An Epidemic of False Claims," *Scientific American* (May 31, 2011),
http://www.scientificamerican.com/article.cfm?id=an-epidemic-of-false-claims (Accessed 6-15-13).

Scientific Contradictions

Today's science is also filled with contradictions. The result of one set of studies is often refuted by the result of another set of studies. Again, nowhere is this more evident than the results of medical studies and the science behind environmental threats to our health. A few questions demonstrate the conflicts in today's science.

Does taking a daily vitamin help maintain good health or is it a complete waste of money? Are eggs good or bad for your health? What's healthier, butter or margarine? Do antibacterial cleaners protect our health or harm our health? Does broccoli prevent cancer or does it contain cancer-causing agents? No matter how you answer these questions, you can find scientific studies supporting your answer.

Hand sanitizers, for example, protect us from transmitted disease, but many also contain carcinogens, hormone disruptors, and ingredients linked to asthma.[17] Even those fruits and vegetables we're told to eat in order to maintain good health contain carcinogens. The list of *dangerous* fruits and vegetables include apples, broccoli, lettuce, peas, corn, potatoes, and others.[18] Is there anything in our environment that won't kill us?

There are two reasons we're not routinely warned about the carcinogens in broccoli. First, the health benefits from eating broccoli far outweigh any risks. At least, that's what science tells us today. The second reason is one we seldom consider. We're not warned about the *dangers* of broccoli because there are no organizations devoted to ridding the world of broccoli. If this seems like a silly reason, it's because, other than its taste, it's hard to imagine anyone dedicated to banning broccoli and because we've grown up learning that broccoli is a health food. Many of us also

---

17. Amanda Green, "9 habits that may do more harm than good," *Mother Nature Network, WomansDay.com* (August 17, 2011), http://www.mnn.com/health/fitness-well-being/stories/9-habits-that-may-do-more-harm-than-good (Accessed 3-12-12).
18. Dr. George Thomas (a.k.a., Robin Motz), "Carcinogens and Vegetables," *Medicine: Facts and Fictions* (May 19, 2009), http://ghthomas.blogspot.com/2009/05/carcinogens-and-vegetables.html (Accessed 3-2-13).

grew up hearing about the benefits of eating eggs, taking vitamins, and washing our hands with hand sanitizers. Are these things still good for us? Is the science complete—and accurate?

The point is our view of broccoli as a health food could easily change. All it would take is one scientific study highlighting the cancer-causing agents in broccoli, a sensationalizing media to broadcast this new threat to our health, and a trusted organization devoted to ridding the world of broccoli. Such a group would ensure that we're repeatedly warned about the dangers of broccoli. Those evil broccoli profiteers would soon go bankrupt and children around the world would rejoice!

In upcoming chapters we'll see a few health scares that were created by little more than a combination of faulty science, a sensationalizing media, and trusted organizations allegedly looking out for our best interests. Our beliefs are easily manipulated.

Why This Matters

This matters because science is increasingly being used to shape our beliefs. It's also becoming an increasingly large part of headline news, which means we'll hear the sensational side of research findings, but we won't hear the limitations or potential biases behind the science. It matters because science has become a tool of persuasion for powerful political lobbies. It matters because science sells, but it also deceives. More importantly, it matters because today's science is often for sale.

Chapter 3
Autism, Science, and the Media

"It's the most appalling catalog and litany of some of the most
terrible behavior in any research . . . ."
—Dr. Richard Horton, editor of *The Lancet*

Today's scientists readily admit they don't know what causes
autism. There are many theories, but none fully explain the alarming
increase in the disease.

Some have suggested the increase may be driven by an
increased awareness, combined with an expanded definition of
autism. As a result, more parents inquire about the disease, more
doctors look for the disease, and a wider range of symptoms fit the
disease. Other theories suggest causes ranging from immune
system disorders and food allergies to atypical brain development.
Some studies have even suggested that children from wealthy
families are more likely to be autistic.[19] Certainly, wealth would be a
surprising medical *problem*. It's likely this finding is an example of
statistics indicating a cause-and-effect relationship that simply
doesn't exist. Such results are not uncommon.

Given the proper motivations, such results can also be
intentionally created with the biased selection of data. It's a common
marketing gimmick. If you cherry-pick the data evaluated, your
results are guaranteed to tell the desired story—whether or not that
story is true.

Perhaps the most recognized theory regarding the cause of
today's increase in autism was developed in a study published in
1998 by Dr. Andrew Wakefield. This study detected a link between

---

19. Lisa Jo Rudy, "Autism at Highest Rate Among Wealthy, White, College
Educated?" *About.com* (January 6, 2010),
http://autism.about.com/b/2010/01/06/autism-at-highest-rate-among-
wealthy-white-college-educated.htm (Accessed 10-7-13).

childhood vaccinations against Measles, Mumps, and Rubella (the MMR vaccine) and an increased probability of developing autism. The results were published in the highly regarded medical journal, *The Lancet*, and received widespread media coverage. The mainstream media was eager to warn of this new *scientifically proven* danger.

Various nonprofits, including many environmental groups, added to the media buzz challenging the safety of childhood vaccinations. One of their primary targets was thimerosal, which is a common vaccine preservative that contains trace amounts of mercury. The thought of injecting children with mercury is scary enough, but when combined with the Wakefield study, it was enough to convince many parents to stop vaccinating their children against easily preventable, deadly diseases. It's a step many parents would have found unthinkable prior to 1998. Needless to say, this study was controversial the day it was released.

If you search the web for "*Lancet* retracts autism study," you'll find that the controversy flared anew in 2010, twelve years after the study's release, when *The Lancet* retracted Dr. Wakefield's paper. The retraction was based on findings from Britain's General Medical Council (GMC) indicating patient selection for the study was biased and the good doctor had been irresponsible and dishonest in his research. Criticism of the study wasn't limited to the GMC.

Alison Singer, president of the Autism Science Foundation and the mother of a child with autism, said the Wakefield study was unscientific and caused significant harm because children were dying from easily preventable deadly diseases. Professor William Schaffner, chairman of the Department of Preventive Medicine at Vanderbilt University, indicated that twenty subsequent government-led studies by various medical researchers in various countries denied any connection between autism and vaccinations.

If you knew of the alleged link between vaccines and autism, but didn't know about these twenty other studies, what does that say about the media's coverage of seemingly critical issues? What does it say about the media's bias for the sensational and their lack of interest in reporting the mundane fundamental facts? The mainstream media may well be the worst source of factual information in existence, yet that's where we routinely turn to learn about our world. It's a formula that leaves us vulnerable to being deceived.

Of course, not everyone agrees with criticizing the Wakefield study. Actress Jenny McCarthy is perhaps the best-known supporter of the link between vaccines and autism. Many others remain convinced that their children had reactions to the vaccine that led to autism. Some have referred to Dr. Wakefield as an honest and courageous doctor who dared to tell an inconvenient truth.

What to Believe?

Was Dr. Wakefield an honest doctor or was he simply seeking his fifteen minutes of fame? Did the GMC do the right thing by finding his study unethical, or did they merely choose the lesser of two evils by accepting the low probability of autism in favor of the known benefits of the vaccine? Even worse, is it possible the GMC acted to protect profits for the vaccine industry? Many believe that's exactly what happened. Such beliefs are reinforced by projections that the global vaccine industry will be grossing $52 billion per year by 2016.[20] Did industry money buy the GMC's decision? It's a common perception and one we apply frequently when governments or regulators make decisions that seemingly favor Big Business.

Then there's the matter of those twenty other government-led studies that found no link between vaccines and autism. Did industry secretly fund those studies? Did governments intentionally bias those studies in hopes of preventing outbreaks of deadly childhood diseases? We generally have little trouble believing in such antics by Big Business and Big Money, but we seldom stop to consider the possible motives of the good guys—our trusted nonprofit watchdog organizations seeking to protect our health by revealing inconvenient truths. More to the point: Did Dr. Wakefield have a motive to bias his study?

---

20. David R Curry, "Global Vaccine Revenues Projected to More than Double by 2016," Center for Vaccine Ethics and Policy (January 17, 2010), http://centerforvaccineethicsandpolicy.net/2010/01/17/global-vaccines-revenues-projected-to-more-than-double-by-2016/ (Accessed 7-15-13).

Unusual Suspects

As it turns out, profit was the motive behind this entire story, but it wasn't industry profit. It was the worst profit of all—ambulance-chasing lawyers. Two years before the study, Dr. Wakefield was added to the payroll of a law firm hoping to sue the vaccine industry.[21] Additionally, one year before releasing his study, Dr. Wakefield applied for a patent on a new measles vaccine that would compete with the MMR vaccine.[22] Regardless of how you or I view the dangers of vaccinations, these motives are appalling. The results are worse: Children are dying from easily preventable diseases and many of these deaths are the direct result of fear created by questionable science and a sensationalizing media.

The Media's Role

The mainstream media gave us the sensational news: Vaccines contain mercury and independent scientific research has linked vaccines to autism. The media spoke the truth. They didn't lie. They also failed to do their homework and failed to tell us the fundamental facts.

They didn't tell us that Wakefield's findings focused on only twelve children. They didn't tell us that five of those twelve children had previously been diagnosed with developmental problems. They never mentioned that timelines were altered in the study to create the illusion of a nonexistent cause-and-effect relationship.[23] They didn't tell us about Dr. Wakefield's hidden motives.

To learn these facts, we had to wait twelve years and turn to the GMC and an investigative report from Brian Deer. The Wakefield study should never have become headline news, but it did. As a result, many parents around the globe have stopped vaccinating their children. Measles outbreaks are reoccurring with a vengeance. Every year 242,000 children die of measles. From 2005

---

21. Brian Deer (of the *Sunday Times of London*), "Secrets of the MMR Vaccine Scare," *British Medical Journal* (BMJ 2011;342:c5347), http://briandeer.com/mmr/lancet-summary.htm (Accessed 3-6-12).
22. Ibid.
23. Ibid.

to 2010 in New York and Connecticut, the number of unvaccinated children doubled. During the same time period, the number of unprotected children in New Jersey increased seven fold. In California, whooping cough has returned to levels not seen since 1958. In July 2013, the *Wall Street Journal* carried a front-page story regarding a measles pandemic that was erupting in Wales. Is our fear of vaccines doing more harm than good?

The media didn't lie to us. They simply failed to tell us the whole story—as they so often fail to do.

Chapter 4
Silent Spring and DDT

". . . [*Silent Spring* is] the most important chronicle of the century for
the human race,"
—US Supreme Court Justice, William O. Douglas

The above quote refers to Rachel Carson's world-changing
1962 nonfiction, *Silent Spring*. This nonfiction work described
humanity's abuse of the environment, the open dumping of
hazardous chemicals, and our haphazard overuse of dangerous
pesticides, especially the pesticide DDT. Carson's book was a much
needed wakeup call. Many consider *Silent Spring* as the catalyst
that started today's environmental movement.

There are two very different views of *Silent Spring* and DDT.
The differences will demonstrate two points of emphasis. First, we
rarely hear both sides of the story. Second, our reactions to
environmental threats often have unintended, negative
consequences—like the reaction of many to Dr. Wakefield's study.

DDT

DDT was first created in 1873, but it wasn't until the late 1930s
that it was synthesized for widespread use as a pesticide. DDT was
a boon to farming after World War II, practically eliminating crop loss
from insects. It also offered the promise of eliminating mosquito-
borne diseases, including deadly malaria. In 1948, the Nobel Prize
in Medicine was awarded to Dr. Paul Muller for his work leading to
the development of DDT as a usable pesticide.

In 1955, the World Health Organization launched a worldwide
program to eradicate malaria. DDT was the primary tool used to
achieve that goal. The program began in developed nations. The
US is now essentially free of malaria as the direct result of this
program and heavy reliance on DDT.

I was very young at the time, but I still remember watching trucks from the County Health Department roll slowly down our street leaving behind a white cloud of DDT mist that drifted through the trees, onto our lawns, over our houses, and into our back yards. I didn't realize it at the time, but I was watching history: the effective end of malaria in the US.

For undeveloped nations, the news wasn't as good. The program to eradicate malaria was stopped in many Third World countries before it was complete. As a result, malaria still kills as many as 3 million people every year. In contrast, the US experiences about 1,500 malaria cases annually and most of those cases are tied to foreign travel.[24]

Why was such a successful lifesaving program stopped short of its goal? The answer lies in the rise of environmentalism that was triggered in 1962 by Rachel Carson's book. *Silent Spring* documented research indicating that DDT could cause cancer, attack the human nervous system, and harm human reproduction. *Silent Spring* highlighted another potential DDT evil. It was thought to cause eagles to produce thin eggshells that would break before the chicks inside were old enough to survive outside the egg. DDT was also thought to affect the fertility rate of some ninety species of birds, including robins. The book's title was a reference to a future world without robins, the cherished songbirds of spring.

The formation of the US EPA was the direct result of the public outcry that followed publication of Carson's book. In what has been hailed as the first major victory for the environmental lobby, on June 14, 1972 the EPA's first administrator, William Ruckelshaus, issued a total ban on the manufacture, sale, and use of DDT. Most nations have implemented similar bans.

As noted at the top of this chapter, US Supreme Court Justice William O. Douglas hailed *Silent Spring* as "the most important chronicle of the century for the human race." Such praise, however, wasn't universal. Many scientists strongly objected to Rachel Carson's findings, citing a number of subsequent studies that refuted key accusations in her book. These scientists also

---

24. US Centers for Disease Control and Prevention, "Malaria and Travels," http://www.cdc.gov/malaria/travelers/ (Accessed 10-7-13).

highlighted the tremendous benefits tied to DDT.

What are we to believe about DDT? Do we believe Rachel Carson or those scientists objecting to her findings? While it's difficult to accuse Rachel Carson of intentionally biasing her findings, we can easily envision numerous motives for those challenging Carson's work. Perhaps the chemical industry funded the scientists refuting Rachel Carson. That's easy enough to believe. We know that numerous doctors and scientists working for Big Tobacco fought the connection between smoking and lung cancer for decades. No one could blame us for believing the same thing was happening with *Silent Spring*.

Today, most would consider the debate over DDT as a thing of the past, just like the debate regarding smoking's impact on lung cancer. That being the case, what are today's scientists telling us about the dangers of DDT?

Many in the scientific and healthcare community are calling the total ban on DDT a gross and tragic overreaction that has been directly responsible for the unnecessary deaths of as many as 60 million victims of malaria. Instead of a total ban, these scientists suggest using controlled applications of DDT to ensure that it protects both the environment and human life. Is it possible these scientists are correct?

The Lies of Rachel Carson

Dr. J. Gordon Edwards was an entomologist who spent years disputing many of the claims made by Rachel Carson. He viewed *Silent Spring* as little more than rogue journalism, filled with exaggerated claims and creative marketing terminology. He notes the frequent use of the terms "poisons" and "agents of death," as well as comparisons of DDT to mustard gas and chemical warfare. In 1992, Dr. Edwards summarized his concerns in a paper titled "The Lies of Rachel Carson."

Dr. Edwards was first introduced to DDT in World War II when he was ordered to spray everyone in his company with DDT to stop a typhus epidemic that was being spread by lice. This action was credited with saving the lives of thousands of troops. Based on his own experience and the health *benefits* provided by DDT, Dr. Edwards began a series of public presentations to refute the findings of *Silent Spring*. At the start of each presentation, he swallowed a tablespoon of DDT to demonstrate that it posed no threat to human health.

In case you're wondering about the health effects of those dramatic demonstrations, Dr. Edwards died of a heart attack in 1994 at the age of 84—while climbing Divide Mountain in Glacier National Park. Heart attacks were not on the list of health risks tied to DDT.

Dr. Edwards also attacked one of the more influential findings of *Silent Spring*: the projected decline in bird populations. Citing the same studies used by Rachel Carson, Dr. Edwards found that after eight weeks of monitoring, the survival rate for chicks from birds overdosed with DDT was higher than the survival rate of chicks from birds that were not fed DDT. In addition, he cited experiments by Joseph Hickey of the University of Wisconsin indicating that robins exposed to DDT simply passed the chemical out of their bodies in their feces—without harm.

To further support his claims, Dr. Edwards pointed to the 1960 Audubon Christmas Bird Count. This count was conducted at the height of DDT usage, yet it showed twelve times more robins per observer than a similar count conducted in 1941.

How Dangerous Is DDT?

In 2002, the US Department of Health and Human Services indicated that people who ingested large amounts of DDT experienced sweating, headache, nausea, vomiting, and dizziness. Some also experienced tremors and seizures. These are terribly frightening reactions that are indicative of the media coverage that surrounded DDT. What weren't as well publicized were the conclusions of this study, which found that these frightening effects "disappeared once exposure stopped."[25] The study indicated that DDT is "reasonably anticipated" to be a human carcinogen at high enough levels of exposure, but also noted that human volunteers

---

25. US Department of Health and Human Services, Agency for Toxic Substances & Disease Registry, "Public Health Statement for DDT, DDE and DDD" (September, 2002), Section 1.5 "How can DDT, DDE, and DDD affect my health?" http://www.atsdr.cdc.gov/phs/phs.asp?id=79&tid=20 (Accessed 11-28-11).

showed no health effects after taking 35 mg capsules of DDT every day for eighteen months.

Referring to the alleged link between DDT and breast cancer, a study at Cornell University indicated that more recent and "better-controlled studies" have failed to identify any such link.[26] Additionally, an article from the Center for Disease Control (CDC) indicated, "Workers heavily exposed to DDT have not had more cancer than workers not exposed to DDT."[27]

In 2004, the Association of American Physicians and Surgeons issued a resolution proposing that agencies of the US and UN allow the use of DDT in countries where malaria is prevalent. The resolution stated that most of today's malaria deaths occur in pregnant women and children under five years of age living in sub-Saharan Africa.[28] Today, an infant dies of malaria every thirty seconds. The resolution indicates that these deaths could be drastically reduced—if not eliminated—by allowing the controlled application of DDT.

How Beneficial Is DDT?

In Sri Lanka, the use of DDT decreased malaria from 2.8 million cases per year in 1948 to a mere seventeen (17) in 1964. Five years after DDT was banned in that country, annual malaria cases

---

26. Suzanne Snedeker (Research Project Leader, Brest Cancer Environmental Risk Factors [BCERF]) and Erica Allgyer, Cornell University, "Pesticides and Breast Cancer Risk: DDT and DDE, Fact Sheet #02" (April 2001),
http://envirocancer.cornell.edu/FactSheet/Pesticide/fs2.ddt.cfm (Accessed 5-10-11).
27. US Department of Health and Human Services, Center for Disease Control and Prevention, "Environmental Hazards and Health Effects Program; Cancer Clusters; Frequently Asked Questions About DDT and DDE, Can DDT and DDE Cause Cancers Including Leukemia?"
http://www.cdc.gov/nceh/clusters/fallon/ddtfaq.htm#cause_leukemia (Accessed 11-30-11).
28. Association of American Physicians and Surgeons, Inc., "Legalizing DDT to Fight Malaria in Tropical Countries," Resolution 61-02, 2004,
http://www.aapsonline.org/resolutions/2004-2.htm (Accessed 4-25-11).

rebounded to 2.5 million cases![29]

There are conflicting stories regarding the dangers of using DDT. Believe whichever side of those stories you want, but one thing is clear: The results of banning DDT have been tragic.

The Bottom Line

It's impossible to overstate the significance of the wakeup call delivered by *Silent Spring*. It's also impossible to ignore the tragic results tied to our decision to ban DDT. We based that decision on a critically important book, the best *independent* science of the day, and the intense media coverage of the alleged dangers. We also made that decision after we were enjoying the luxury of living in a malaria-free world. Would we have made the same decision if malaria was killing an infant every thirty seconds in our country?

---

29. Todd Seavey, "The DDT Ban Turns 30: Millions Dead of Malaria Because of Ban, More Deaths Likely" (June 1, 2002), American Council on Science and Health,
http://ruby.fgcu.edu/courses/Twimberley/10199/DDTPaper.pdf   (Accessed 4-25-11).

Part II
Surprising Deceptions and Deceivers

"Ultimately, a trend toward abandoning science objectivity in favor of political agendas forced me to leave Greenpeace."
—Dr. Patrick Moore, former Director at Greenpeace

Chapter 5
Asthma and the Best Kept Secret

". . . indoor levels of pollutants may be 2 to 5 times—and occasionally more than 100 times—higher than outdoor pollutant levels."
—US EPA[30]

Imagine holding your breath underwater until you're about to burst, then springing to the surface only to find you're still unable to breathe. Asthma has to be a frightening disease. Like autism, the number of people diagnosed with asthma is rapidly increasing. In the US, diagnosed cases of asthma have increased 75% in the last twenty years. Among preschoolers, that increase is 160% and the largest increase has occurred among inner-city children.[31]

Also like autism, the exact cause of asthma is unknown, but many new theories are emerging to explain the recent increase.[32] These emerging theories are important for two reasons: 1) They coincide with society-wide behavioral changes over the past twenty or more years, and 2) They have nothing to do with industrial air pollution.

Search the web for "genes linked to asthma" and you'll find that genetics are increasingly thought to play a key role in developing

---

30. US EPA, "Questions About Your Community: Indoor Air," http://www.epa.gov/region1/communities/indoorair.html (Accessed 6-15-13).
31. Jeanie Lerche Davis, "Asthma Cases on the Rise," *WebMD Health News* (May 5, 2000), http://www.webmd.com/news/20000505/asthma-cases-on-rise (Accessed, 3-27-09).
32. US Department of Health and Human Services; National Heart, Lung, and Blood Institute, "What Causes Asthma," http://www.nhlbi.nih.gov/health/health-topics/topics/asthma/causes.html (3-27-09).

asthma. Under this theory, if you don't inherit the required genes from your parents, environmental exposures won't cause you to develop asthma.

Another theory contends that children who don't get enough exercise are more likely to suffer from asthma.[33] Under this theory, the increase in obesity and increasingly sedentary lifestyles may be linked to the increase in asthma. Of course, there's also exercise-induced asthma, which means asthma can be triggered by too much exercise. What are we to do?

A different theory suggests that our homes are actually too clean and sanitary.[34] This theory contends that children in Western countries are no longer exposed to many of the early childhood infections that were common just a generation ago. As a result, children's immune systems don't develop a robust response to common allergens, which leaves them vulnerable to asthma. We should also remember that many of today's antibacterial soaps have been linked to asthma. If you suffer from asthma, perhaps you can blame mom for keeping a clean house and insisting that you always wash your hands.

While we're on the subject of mom, another theory concludes that infants who aren't breastfed for at least their first six months of life are more likely to get asthma.[35] Today, fewer than half of all US babies are breastfed for the recommended length of time—

---

33. *MedicineNet.com*, "Exercise Preventing Asthma" (February 15, 2007), http://www.medicinenet.com/script/main/art.asp?articlekey=16696 (Accessed 9-25-11).

34. US FDA, Department of Health and Human Services; "Vaccines, Blood and Biologics: Asthma, the Hygiene Hypothesis," http://www.fda.gov/biologicsbloodvaccines/resourcesforyou/consumers/ucm167471.htm (Accessed 1-9-13).

35. Agnes M.M. Sonnenschein-van der Voort, Vincent W.V. Jaddoe, Ralf J.P. van der Valk, Sten P. Willemsen, Albert Hofman, Henriëtte A. Moll, Johan C. de Jongste, and Liesbeth Duijts; "Breastfeeding May Prevent Asthma" (July 23, 2011), European Lung Foundation, http://www.european-lung-foundation.org/17433-breastfeeding-may-prevent-asthma.htm (Accessed 4-30-13).

assuming this theory is correct.[36]

People with existing allergies are also more likely to develop asthma than those without allergies.[37] The allergens most often tied to asthma include cockroaches, dust mites, second-hand smoke, pet dander, mold, and pollen.[38] With the possible exception of pollen, our greatest exposure to these allergens and asthma triggers actually occurs indoors. Is it possible our health is at greater risk from indoor air quality rather than outdoor air quality?

According to the EPA:

> A growing body of scientific evidence has indicated that the air within homes and other buildings can be more seriously polluted than the outdoor air in even the largest and most industrialized cities. Other research indicates that people spend approximately 90 percent of their time indoors. Thus, for many people, the risks to health may be greater due to exposure to air pollution indoors rather than outdoors. . . . If too little outdoor air enters a home, pollutants can accumulate to levels that can pose health and comfort problems. Symptoms of some diseases, including asthma . . . may also show up soon after exposure to some indoor air pollutants.[39]

> *Note: The strong warning shown above has been removed from the referenced EPA website (sometime between July

36. Jerry Norton (*Reuters*, September 13, 2010), "Fewer than half of US mothers breastfeed enough: CDC",
http://www.reuters.com/article/2010/09/13/us-breastfeeding-usa-idUSTRE68C47D20100913 (Accessed 4-30-13).
37. Elana Pearl Ben-Joseph, "Do Allergies Cause Asthma?" (October, 2010), *KidsHealth*,
http://kidshealth.org/parent/asthma_center/asthma_allergies/allergies_asthma.html (Accessed 4-30-13).
38. Melinda Ratini, "Allergies and Asthma," *WebMD, Asthma Health Center* (July 10, 2012), http://www.webmd.com/asthma/guide/allergies-asthma (Accessed 4-30-13).
39. US EPA, "An Introduction to Indoor Air Quality (IAQ)," http://www.epa.gov/iaq/ia-intro.html (Accessed 7-23-10).

23, 2010 and May 1, 2013). It was replaced with softer language, but the softer language still warns that the effects of indoor air quality can cause "respiratory diseases, heart disease, and cancer."

Common sources of harmful indoor pollutants include cleaners, aerosols, tobacco, fireplaces, cooking, air fresheners, cockroaches, pets, human skin, dust, dust mites, paint, carpet, radon, mold, and a host of other things routinely found in homes and offices. The EPA suggests one way to lower concentrations of indoor pollution is to increase the amount of outdoor air coming indoors.[40] According to Livestrong.com, a website dedicated to healthy living, "Because of the heightened concentration levels, the worst exposure to carcinogens and dangerous pollutants tends to occur in indoor environments [rather than outdoors]."[41]

Why This Is Important

All of the above theories regarding asthma match changes in lifestyles that have evolved across the same timeline as the rapid increase in asthma. These theories also identify indoor air quality as the primary concern for our health, not outdoor air quality. More importantly, none of these theories point to industrial air pollution, which clearly doesn't match the popular perception.

Because asthma is a respiratory disease with symptoms that are triggered by airborne irritants, we commonly associate increased asthma with increased industrial air pollution. It's a seemingly obvious conclusion and one that's routinely reinforced by the media and environmental organizations. There's a major problem with this perception. Not only is it wrong, it's actually impossible to link the increase in asthma to increased industrial air pollution!

---

40. US EPA, "An Introduction to Indoor Air Quality (IAQ), Improving Indoor Air Quality," http://www.epa.gov/iaq/is-imprv.html (Accessed 12-3-12).
41. Jacob S, "Outdoor Air vs. Indoor Air," *Livestrong.com* (July 26, 2010), http://www.livestrong.com/article/185502-outdoor-air-vs-indoor-air/ (Accessed 10-12-11).

The Best Kept Secret

We can't blame pollution from industry for the increase in asthma because US pollution from those sources has steadily and continually decreased every year since 1970! In fact, the total volume of critical pollutants emitted today is lower than it's been for decades! Hard to believe, isn't it?

During this same period in the US, the population increased 52%, the number of annual miles driven more than doubled, the economy tripled, and energy consumption increased 48%. Over the past 40-years, we've increased energy use nearly 50%, yet we've drastically reduced critical emissions from its use. That's clearly not the message we've heard from the mainstream media or the environmental lobby.

Since 1970, the EPA reports a 68% decrease in the amount of Criteria Pollutants emitted in the US![42] The Criteria Pollutants include sulfur dioxide (SO_2), nitrogen oxides (NOx), lead (Pb), particulate matter (PM), ground-level ozone (smog), and carbon monoxide (CO). We should note that ozone is not a direct emission. Ozone—a.k.a. smog—is formed during hot, calm weather when NOx and volatile organic compounds (VOCs) concentrate at ground level and are adequately cooked by the sun.

Between 1980 and 2010, the EPA cites the following reductions in overall emissions: Carbon monoxide emissions dropped 71%; lead, 97%; nitrogen oxides, 52%; and sulfur dioxide, 76%.[43] Additionally, emissions of VOCs dropped 63%; Ozone air quality improved 28%; and the amount of particulate emissions dropped 55% for very small, fine particulates (PM 2.5) and 83% for slightly larger, fine particulates (PM 10).[44]

These reductions are shown graphically in Figure 5-1 on the next page.

---

42. www.epa.gov/airtrends/images/comparison70.jpg (Accessed 5-1-12).
43. US EPA, "Air Quality Trends,"
http://www.epa.gov/airtrends/aqtrends.html (Accessed 2-15-13).
44. Ibid.

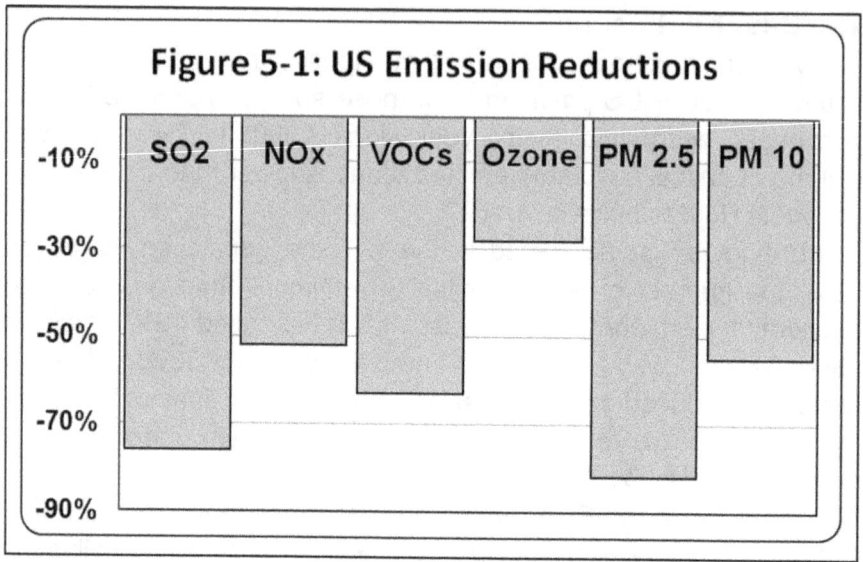

Figure 5-1: US Emission Reductions

This has to be the best-kept secret in history.

Impossible to Believe

If you find the reduction in air pollution hard to believe, you're not alone—and that's the point. Literally everything we're told about the state of the environment paints a doomsday picture. We're told that energy use and asthma are on the rise, so there must be a cause-and-effect relationship. After all, the statistics prove the obvious connection, don't they? Obviously, they don't.

Some might question the possible cumulative impact of decade after decade of ongoing air pollution. The answer to this challenge can be found in the product names of many air fresheners like *April Fresh* and *Spring Rain*. Rain washes irritants out of our skies, helping to erase the potential cumulative effect of most pollutants, especially asthma triggers.

Yes, it's hard to believe pollution has declined and our air quality has improved. However, if we stop long enough to consider the fundamental changes over the past thirty to forty years, we'll realize that cleaner air is exactly what we should expect. Today, we build cars that are much cleaner and more efficient than cars of the past, we produce gasoline that is cleaner, and we no longer burn trash in our back yards. Industry has also spent hundreds of billions of dollars cleaning up dirty smokestacks. These fundamentals reinforce the above EPA findings. It's just hard to believe our air is cleaner—and even harder to find anyone to tell us about it.

About Las Vegas

I have one friend in particular who refuses to believe industrial pollution has declined. He's adamant that air quality is worse today than ever before and he points to Las Vegas as real-world, tangible proof. Some thirty-plus years ago when he first visited Clark County, Nevada, the home of Sin City, the air was clear and pristine. Today when he visits, he sees a brown haze on the horizon and his asthma symptoms flare soon after arriving.

If you live in a growing city, you've likely experienced this same deteriorating air quality at home. You've seen it with your own eyes, so you have proof that air quality is getting worse. How then, do we reconcile such personal experiences with the EPA data regarding decreased pollution?

The answer boils down to concentrated population growth. In other words, overall air quality has improved, but cities with booming populations are likely experiencing deteriorating air quality. Las Vegas serves as a prime example.

In 1980, about the time of my friend's first visit to Sin City, Clark County's population was roughly half a million; by the year 2000, the population had tripled to more than 1.5 million; and the 2010 census indicated that some 1.9 million citizens lived in and around Las Vegas—not counting tourists.

Las Vegas sits on the desert floor surrounded by mountains. Unless a weather front blows through, the pollutants emitted in Vegas are trapped. When it comes to pollution, the old expression is true: "What happens in Vegas stays in Vegas." That leads to an obvious question. What sources are responsible for pollution in Las Vegas? Is it industry, or the exploding local population? To find out, we can turn to an environmental scorecard that examines pollution in the US.

Sources of Pollution

The GoodGuide Scorecard is a website run by a team working for the Environmental Defense Fund.[45] At this website, you can plug in your zip code and find the sources of pollution in any region of the US. The Scorecard shows emission levels from three different sources: Mobile sources, Area sources, and Point sources. Mobile sources include planes, trains, and automobiles. Area sources include homes, offices, strip malls, casinos, lawnmowers, barbeque grills, gas stations, dry cleaners, and similar facilities. Point sources are large stationary emitters, such as coal-fired power plants and industrial facilities.

For Clark County, Nevada, the scorecard shows that Mobile and Area sources are responsible for over 66% of all nitrogen oxides emissions and a full 100% of all VOCs: the two primary components of smog (ground-level ozone). Mobile and Area sources are also responsible for emitting 88% of all Criteria Pollutants combined.[46] Such results aren't limited to Sin City.

In New York City, New York, Mobile and Area sources are responsible for emitting 87% of all nitrogen oxides, 99% of all VOCs, and 96% of all Criteria Pollutants combined.[47] In Los Angeles, California, Mobile and Area sources are responsible for emitting 93% of both nitrogen oxides and VOCs while emitting 97% of all Criteria Pollutants combined.[48] Even in more industrialized cities like Cleveland, Ohio, Mobile and Area sources are the largest emitters

---

45. *Goodguide Scorecard*, "Pollution in Your Community, Update on Scorecard," http://scorecard.goodguide.com/ (Accessed 3-15-13).

46 .Scorecard by the GoodGuide, http://scorecard.goodguide.com/env-releases/cap/county.tcl?fips_county_code=32003#emissions_summary (Accessed 9-10-11).

47. Scorecard by the GoodGuide, http://scorecard.goodguide.com/env-releases/cap/county.tcl?fips_county_code=36061#emissions_summary. (Accessed 9-10-11).

48. Scorecard by the GoodGuide, http://scorecard.goodguide.com/env-releases/cap/county.tcl?fips_county_code=06037#emissions_summary (Accessed 10-13-11).

of all Criteria Pollutants with the lone exception of sulfur dioxide.[49]

Unfortunately, the data provided by the GoodGuide is from 1999 and sadly outdated. Fortunately, the EPA provides annual, nation-wide emission summaries that tell us how emission levels have changed since that time. Between 1999 and today (2014), emissions of all Criteria Pollutants decreased, but Point source emissions (industrial emissions) declined more than emissions from Mobile or Area sources. Area source emissions actually *increased* for all Criteria Pollutants except small particulates. As shown by the GoodGuide, Mobile and Area sources were and remain the source of over 88% of all Criteria emissions in Las Vegas, New York, and Los Angeles. The fact is cars, homes, and small businesses combine to be the greatest sources of pollution in large cities, the areas suffering our most polluted skies.

Does Location Matter?

We're told that the location of an emission source is irrelevant because pollution from remote industries is transported to major cities where it mixes with local pollution to create the air quality problems experienced in those locations. Although this argument makes sense, the fundamental facts cast doubt on the validity and significance of transported pollution.

Smog (ground-level ozone) is at its worst when temperature inversions trap local pollution. In other words, air quality becomes a problem when winds are calm. Much like Las Vegas sitting in a bowl, air quality in our largest cities improves when winds push local pollution out of the area—and bring pollution from remote regions into our cities. Our continued focus on industrial pollution is costing jobs, increasing the cost of energy, increasing the cost of consumer goods, and most importantly, failing to address the primary sources of our filthiest air. Are we certain our continual focus on industrial pollution is the right focus?

A quick review of air quality in California provides the answer.

---

49. Scorecard by the GoodGuide, http://scorecard.goodguide.com/env-releases/cap/county.tcl?fips_county_code=39035#emissions_summary (Accessed 10-13-11).

Randall L Hughes

Ozone and California

The EPA produces a map of US counties violating ozone standards. Not surprisingly, the map primarily highlights major metropolitan areas.[50] It also highlights roughly two-thirds of the state of California.

California has arguably the toughest environmental regulations in the US. They've hammered away at industrial pollution for decades, yet according to the American Lung Association's 2011 annual State Of The Air Report, eight of the ten worst cities for ozone pollution were located in California. The state also held the top two spots as the worst cities for fine particulate pollution.[51]

The news regarding California's air quality isn't good, but Californians can't blame industry or manufacturing because the state's tough environmental regulations have driven many of those job-creating businesses to other states—we'll see details in later chapters. California also can't blame pollution transported from Midwest industries because the Pacific breezes that dominate regional wind patterns can't transport pollution from the Midwest across the Rockies to the West Coast. Finally, Californians can't blame their coal-fired power plants because nearly all their coal plants have closed. As far back as 2005, in-state coal plants provided less than 1% of the state's total electric generating capacity—and no new coal plants have been built in California since then.[52]

Consequently, California's air quality problems are caused by the same sources creating problems in Las Vegas, New York, and other major cities: Area sources and cars. The fundamental reality is that our most polluted skies are caused by rush hour traffic and the concentration of homes, apartments, and office buildings—not

---

50. US EPA, "8-Hour Ozone Nonattainment Areas (1997 Standard)" (December, 2012), http://www.epa.gov/oaqps001/greenbk/map8hr.html (Accessed 3-12-13).

51. American Lung Association, "State of the Air 2011, Most Polluted Cities," http://www.stateoftheair.org/2011/city-rankings/most-polluted-cities.html (Accessed 9-15-12).

52. Source Watch, "California and Coal," http://www.sourcewatch.org/index.php?title=California_and_coal (Accessed 3-9-10).

industry.

Attacking the Wrong Problem

Like California, we've hammered away at industrial pollution for decades. Our reward has been improved air quality everywhere except where it's needed most: our major cities. In terms of reducing industrial pollution, we've picked the low-hanging fruit, meaning we've cleaned our most prolific sources of industrial emissions. Future reductions from industry will cost increasingly more while yielding ever-smaller health-related improvements. We're attacking the wrong problem.

As previously noted, the largest increase in asthma has occurred among inner-city children, where Mobile and Area sources are responsible for the vast majority of Criteria Pollutants, and where we spend 90% of our time indoors. Most of us continue to view industry as the root of all environmental evil. Are we certain this perception is accurate?

The Ozone Map and Future Perceptions

When a city is highlighted on the EPA's map of ozone violators, it's common to assume that the air in that city is dangerous all the time. That's not the case. In fact, it takes as little as thirty-two hours of poor air quality per year for a city to violate EPA ozone standards. That's all the time it takes for the media to tell us we live in regions with dangerous air quality. As a result, our perception of air quality is not good—and it's about to get much worse.

The EPA has announced tougher new air quality standards, including significantly tougher new ozone standards. When these new standards take effect, it's projected that 565 new counties will be added to the EPA's map of regions failing to meet ozone standards.[53] Air quality in these 565 counties may not change—it might actually improve—but these 565 counties will soon fail to

---

53. Paul Bedard, "New Fears, EPA Smog Rule Will Cost 7.3 Million Jobs," *US News* (July 19, 2011),
http://www.usnews.com/news/washington-whispers/articles/2011/07/19/new- fears-epa-smog-rule-will-cost-73-million-jobs-new-fears-epa-smog-rule-will-cost-73-million-jobs (Accessed 1-13-13).

meet the new standards. We'll likely be told that more of us than ever before are living with dangerous air quality. We'll be convinced that industrial pollution must be stopped at all costs. We'll continue to attack the wrong problem even though industrial emissions and ozone levels have steadily and continuously declined for decades.[54]

The Cost

We attack industry because that's the easy target. We don't like the obscene profits of Big Business, and we hate the even more obscene salaries and bonuses given to top executives. We target industry because we want to make the scoundrels pay! Of course, they're not the ones ultimately paying.

When a city or region violates EPA ozone standards, the cost of living and the cost of creating jobs in that region increases. The increased costs are driven by mandated actions required to lower ozone-causing emissions. Businesses face a host of new costs for environmental retrofits or other actions to lower their ozone footprint. Those costs are ultimately passed on to consumers. Costlier vehicle inspections are often required. Cleaner burning, higher priced gasoline is often required on a region-wide basis, as well. One of the most costly ozone-reducing actions deals directly with job creation. Before anyone can add a new business or expand an existing business, they'll have to *turn off* an existing local source of pollution that is equal to or greater than the added pollution they might create. In many cases, the region can't add new jobs unless they lose existing jobs.

The costs and actions required to comply with the new ozone standard alone is projected to destroy 7.3 million jobs and cost the nation 1 trillion dollars by 2020.[55] These aren't the numbers we'll hear from the mainstream media and certainly not the message

---

54. US EPA, "Ozone," http://www.epa.gov/airtrends/ozone.html (Accessed 1-13-13).
55. Paul Bedard, "New Fears, EPA Smog Rule Will Cost 7.3 Million Jobs," *U.S. News* (July 19, 2011),
http://www.usnews.com/news/washington-whispers/articles/2011/07/19/new-fears-epa-smog-rule-will-cost-73-million-jobs-new-fears-epa-smog-rule-will-cost-73-million-jobs (Accessed 1-13-13).

we'll hear from the environmental lobby. Instead, we'll likely hear about the tremendous health improvement we'll experience as the direct result of the EPA's tougher new regulations—even if the air quality where we live remains unchanged.

Over time, these tougher new standards will lead to improved air quality in our largest cities. Since we all want to achieve that goal, why would anyone complain about these new standards? There are numerous reasons: It's a shotgun approach that costs more and accomplishes less than it should; it costs jobs; and it may ultimately do more to harm our health than improve it. This last reason is no doubt controversial. How can cleaner air possibly harm anyone's health?

The answer is easier to defend than you might think, but it's an answer we seldom consider. When the cost of achieving cleaner air becomes too high, standards of living decline and human health ultimately suffers. Korea provides a tangible, real-world demonstration reinforcing this controversial claim.

Chapter 6
Lessons from Korea

". . . economic growth-development and inexpensive energy drive
worldwide improvements in health and longevity. Unnecessarily
stringent regulation tends to depress economic growth and standard
of living, leading to negative health effects."
—The Annapolis Center for Science-Based Public Policy[56]

We're fortunate. We have the luxury of taking for granted the
multitude of health benefits tied to a robust economy. We also take
for granted the critical role that low-cost energy plays in helping to
maintain a robust economy.

Read any report on the subject from the World Health
Organization (WHO) or the UN Development Program and one
theme is clear—Access to affordable energy is essential for
increased economic development, improved standards of living,
improved human health, and increased longevity. Such statements
by the U.N. are generally directed toward undeveloped nations, but
they hold true in developed nations, as well. The reverse also holds
true. When regulations increase the cost of energy, they not only
increase the cost of doing business and the price of consumer
goods; they also slow job growth, reduce standards of living, and
harm human health. Granted, this is not the generally held view of
regulation. Regulations are in place to protect us, not harm us. If
that's the case, how can anyone say that regulation is a bad thing?
A quick look at Korea provides an answer.

About Korea

Prior to 1948, Korea was one country. The economy, the

---

56. The Annapolis Center for Science-Based Public Policy, "Economic
Growth and Low cost Energy Drive Improved Public Health" (2006).

culture, and the people of Korea were united. The level of energy usage, pollution, and public health were essentially the same because there was only one nation: Korea. That changed in 1948 when the UN split the country in half along the 38th parallel. After the split, the two countries developed radically different economies, different energy infrastructures, and different sources of pollution. The South Korean economy thrived while North Korea's economy crashed.

To drive their thriving economy, South Koreans burn a lot of fossil fuel. South Korea burns 135 times more oil, 34 billion cubic feet more natural gas, and 100 million more tons of coal per year than North Korea[57]—and they live thirteen years longer than their relatives in the North.[58] The average South Korean also grows one to three inches taller than the average North Korean.[59]

Average per capita CO_2 emissions in South Korea are nearly seven times greater than per capita emissions in the north. In 2006, *Air Pollution News* indicated that air quality in South Korean cities was the worst among all OECD nations.[60] The OECD is the Organization for Economic Cooperation and Development, which is comprised of more than thirty of the world's most developed nations

---

57. US Central Intelligence Agency, "The World FactBook;" https://www.cia.gov/library/publications/the-world-factbook/wfbExt/region_eas.html (Accessed 5-12-11); https://www.cia.gov/library/publications/the-world-factbook/geos/kn.html; and https://www.cia.gov/library/publications/the-world-factbook/geos/ks.html (Accessed 5-12-11).
58. The World Bank, "World Development Indicators, Life Expectancy," http://www.google.com/publicdata?ds=wb-wdi&met=sp_dyn_le00_in&idim=country:KOR&dl=en&hl=en&q=life+expectancy+in+korea#ctype=l&strail=false&nselm=h&met_y=sp_dyn_le00_in&scale_y=lin&ind_y=false&rdim=country&idim=country:KOR:PRK&hl=en&dl=en (Accessed 5-12-11).
59. *Target Map*, "Average Height by Country," http://www.targetmap.com/viewer.aspx?reportId=28572 (Accessed 11-12-12).
60. *Air Pollution News*, "South Korea air pollution worst among OECD nations" (September 26, 2006), http://www.airpollutionnews.com/south-korea-air-pollution-worst-among-oecd-nations/ (Accessed 7-24-12).

including the US and the UK. Despite such polluted skies, South Koreans also live longer than Americans and the British.[61]

If we believe what we're told by the media and the environmental lobby regarding pollution, South Koreans have no chance for a long, healthy life, yet they outlive North Koreans by a wide margin, enjoy a far superior quality of life, a higher standard of living, and far better health. How are these results possible in the most polluted cities of any OECD country when thousands of independent medical studies conclude that pollution causes numerous health problems and premature death?

These results are possible because all of those medical studies do the same thing we all do. They take for granted the tremendous health benefits tied to a strong economy and they ignore the critical link between low-cost energy and economic growth. These medical studies assume that the only change being made is a reduction in pollution. They never consider the cost of achieving those reductions or the cascading, detrimental economic impacts that may result. They also completely ignore the negative health impacts of lost jobs and a slowing economy.

We've always assumed that the cost of reducing pollution was low and the health benefits were high. That may have been true in past decades, but we've already cleaned our worst sources of pollution—and our air is far cleaner than South Korea's. We've cleaned the air to the point where indoor air quality is routinely the most harmful air we breathe. As previously mentioned, future environmental improvements from industry will cost increasingly more while providing ever-smaller benefits. The cost of reducing pollution increases the cost of doing business, increases consumer prices, and makes it more difficult to compete for jobs in the world market. When taken to the extreme, this lowers standards of living to the point that human health suffers. This is especially true if we're targeting the wrong sources of pollution.

With today's intense global competition for jobs, is it possible we've reached a point of diminished returns? Have we reached the point where the economic toll of added regulation on industrial

---

61. The World Bank, "Life expectancy at birth, total (years)," http://data.worldbank.org/indicator/SP.DYN.LE00.IN. (Accessed 5-12-11).

pollution does more to harm human health than protect it?

Premature Death

The CDC reports that lifestyles and behaviors account for seven out of every ten deaths in the US. They also advise that 70% of all US healthcare costs are related to diseases that can be prevented or controlled by simple lifestyle changes.[62] We blame industry, chemicals, and pollution for harming our health. If we look beneath the sensational headlines, however, we might realize that of all the things that harm our health or cause premature death, industrial air pollution may be the least of our worries.

Consider the health warnings tied to simply grilling dinner. When fat and protein meet high temperatures, all kinds of toxic chemicals are released.[63] Many of these toxins are linked to diabetes and cardiovascular disease. Other toxins have been linked to cancer of the lungs, bladder, prostate, pancreas, and colon. A study of 3,000 women reported a 47% increased risk of breast cancer from a diet high in fatty proteins grilled at high temperature.[64] Who knew that 4th of July cookouts were so deadly?

A web search for "causes of premature death" reveals causes we rarely consider, including too much sleep, too little sleep, too much exercise, not enough exercise, undercooked meat, overcooked meat, not taking vitamins, taking too many vitamins, post-traumatic stress disorder, hot dogs, stress, gum disease, watching TV, low fruit and vegetable intake, sitting, bipolar disorder, not being able to afford air conditioning, sun tanning, airline travel, and lack of health insurance—to name just a few.

The impact of not having health insurance alone is startling. In

---

62. US Centers for Disease Control, "Indicators for Chronic Disease Surveillance" (September 10, 2004),
http://www.cdc.gov/mmwr/preview/mmwrhtml/rr5311a1.htm (Accessed 3-15-13).
63. Laura Schwecherl, "Is Grilling Bad for Your Health,"
TheActiveTimes.com (May 30, 2013),
http://xfinity.comcast.net/blogs/lifestyle/2013/05/30/is-grilling-bad-for-your-health/ (Accessed 8-7-13).
64. Ibid.

1993, a study published in the *Journal of American Medicine* found that the uninsured aged 25 to 74 faced a 25% higher risk of dying than those with insurance.[65] The Institute of Medicine found that 18,000 died prematurely in 2000 because they didn't have health insurance.[66] The Urban Institute found that 137,000 died prematurely between 2000 and 2006 due to lack of health insurance. Twenty-two thousand of those deaths occurred in 2006 alone.[67] According to *FactCheck.org*, a study by Harvard researchers indicated that the uninsured aged 18 to 64 were 40% more likely to die prematurely than those with health insurance.[68] What happens to your ability to afford health insurance if you lose your job or the cost of living increases such that you can't afford to visit the doctor?

The list of things that cause premature death begs a question. Is it no longer possible for people to simply wear out and die of old age? Have natural causes stopped being a cause of death? It's as though we have to find some human activity to blame for every health problem in existence—and pollution is our favorite answer. If we're looking for something to blame, perhaps we should consider falling aircraft parts.

In the US, your chance of dying from falling aircraft parts is

---

65. Institute of Medicine, Staff Report, "Care without coverage: too little, too late" (May 2002) 1, http://www.iom.edu/~/media/Files/Report%20Files/2003/Care-Without-Coverage-Too-Little-Too-Late/Uninsured2FINAL.pdf (Accessed 5-4-13).
66. Ibid, 163.
67. Stan Dorn, "Uninsured and dying because of it," *The Urban Institute* (January, 2008), Summary, 1, http://www.urban.org/UploadedPDF/411588_uninsured_dying.pdf (Accessed 5-4-13).
68. *Fact Check.org*, "Dying from Lack of Insurance" (September 24, 2009), http://www.factcheck.org/2009/09/dying-from-lack-of-insurance/ (Accessed 1-10-13).

thirty-times greater than your chance of dying from a shark attack.[69] Perhaps the EPA should require hard hats be worn at all times?

Beyond this threat to our health, we should also consider deadly hyponatremia, which can cause death due to exposure to the chemical compound known as di-hydrogen monoxide. Di-hydrogen monoxide has been linked to numerous deaths, yet it is heavily utilized in households and industries around the globe. More alarmingly, this deadly compound remains a key ingredient in popular soft drinks and adult beverages. Young women are twenty-five times more likely to die from this syndrome than men of any age, yet we do nothing to protect women from this deadly condition.[70]

If you're a fan of Penn and Teller, you already know that di-hydrogen monoxide is two hydrogen atoms combined with one atom of oxygen, a.k.a. H_2O or water. With hyponatremia, death results from drinking so much water that it short-circuits the body's chemistry. Penn and Teller produced a video spoof in which average citizens were asked to sign a petition banning deadly di-hydrogen monoxide. Not surprisingly, nearly everyone who heard the dire statistics agreed to sign the petition.

The Penn and Teller video shows how eager we are to accept bad environmental news, how easily we can be deceived, and how important it is to look for the fundamental facts before we react.

For air quality, the fundamental facts tell us that industrial pollution is on the decline, indoor air quality is routinely the most dangerous air we breathe, and our worst outdoor air quality is caused by Mobile and non-industrial sources of pollution. If we continue to attack industrial pollution—while losing jobs in the process—are we protecting our health or are we actually harming it?

---

69. Daniel Kahneman, "11. Judgment and Decision Making," Annenberg Learner (*Learner.org*), Adapted from "The Psychology of Judgment and Decision Making," by Scott Plous, McGraw-Hill Higher Education (1993), http://www.learner.org/discoveringpsychology/11/e11expand.html (Accessed 5-2-13).

70. Bob Murray, "Hyponatremia In Athletes," Gatorade Sports Science Institute (GSSI), Sports Science News, http://www.uni.edu/dolgener/UG_Sport_Nutrition/Articles/ssn_hyponatremia.pdf (Accessed 5-6-13).

Lessons from Korea

Obviously, we want clean air, we want to rid our cities of smog, and it's imperative that we protect the environment. We want all of these things, but do we ever stop to consider the costs involved and the resulting negative economic impacts that may ultimately do more to damage our health than improve it? The universal answer is "no."

The mainstream media, the environmental lobby, and the entire healthcare industry tell us our health is compromised, our quality of life is diminished, and our lives are shortened because we use fossil fuels that pollute the air. Korea offers a stark contrast to this popular line of thinking. As vividly demonstrated in a real-world setting, the strong economy of South Korea greatly improves the health of South Koreans despite the worst air pollution of any OECD nation. Conversely, the dire poverty of the North wrecks the health of North Koreans.

Somewhere between these two extremes lies a balance point where new environmental regulations begin to do more harm than good. Have the US and Europe reached that balance point?

Looking at the numbers, environmentalists would have to agree that North Koreans are living a more sustainable lifestyle. In the North, cars are a luxury that few can afford, many till their fields with animal and human labor, and the per capita carbon footprint is seven times smaller than the per capita carbon footprint of South Korea. North Koreans are living the environmental dream!

Given the choice, where would you rather live?

Chapter 7
Irrational Environmentalism

"Ultimately, a trend toward abandoning science objectivity
in favor of political agendas forced me to leave Greenpeace."
—Dr. Patrick Moore, former director of Greenpeace

Seven billion of us now inhabit planet Earth. We can translate that to mean protecting the environment is a necessary passion and our greatest responsibility. That responsibility, however, also requires that our actions be well reasoned, unbiased, and based on an understanding of all the facts. Unfortunately, when it comes to critical issues related to energy, health, and the environment, we seldom hear all the facts. More often than not, what we do hear is sensationalized beyond reason and often hides the fundamental truth. You don't have to take my word for it. Listen to the words of one of the greenest people on the planet: Dr. Patrick Moore.

Why Dr. Moore Left Greenpeace
After completing his PhD in Ecology in 1974, Dr. Moore became one of the five original directors of Greenpeace. He was obviously a strong environmental advocate, but his views of the environmental lobby have changed. On Earth Day 2008, the *Wall Street Journal* carried an article written by Dr. Moore titled "Why I Left Greenpeace."

In the article, Dr. Moore described a change in the leadership and direction of Greenpeace. Instead of focusing on the unbiased science behind environmental issues, he indicated the other directors "were either political advocates or environmental entrepreneurs."

Greenpeace was becoming a political organization that ignored science to promote its version of environmentalism. This change at Greenpeace convinced Dr. Moore to leave that organization.

Dr. Moore began his article by explaining that Greenpeace and other environmental groups opposed chlorine in drinking water (not fluoride, although that too is a target for environmentalists). This opposition existed even though, according to Dr. Moore:

> Science shows that adding chlorine to drinking water was the biggest advance in the history of public health, virtually eradicating water-borne diseases such as cholera. And the majority of our pharmaceuticals are based on chlorine chemistry. Simply put, chlorine is essential for our health. My former colleagues ignored science and supported the ban, forcing my departure. Despite science concluding no known health risks—and ample benefits—from chlorine in drinking water, Greenpeace and other environmental groups have opposed its use for more than 20 years.

Chlorine has been studied time and again by governments around the globe with no findings of health hazards tied to consuming chlorinated drinking water—and countless benefits. The environmental lobby knows of these findings, yet they oppose "the biggest advance in the history of public health." It's a position that seems irrational.

On a more current topic, Dr. Moore discusses the environmental lobby's attack on phthalates, the compounds that make plastic flexible. In the article, he defends the safety of phthalates by saying:

> They [phthalates] are found in everything from hospital equipment such as IV bags and tubes to children's toys and shower curtains. They are among the most practical chemical compounds in existence. Phthalates are the new bogeyman. These chemicals make easy targets since they are hard to understand and difficult to pronounce. Commonly used phthalates, such as diisononyl phthalate (DINP), have been used in everyday products for decades with no evidence of human harm. DINP is the primary plasticizer used in toys. It has been tested by multiple government and independent evaluators, and found to be safe.

Despite such findings, environmentalists have unleashed a marketing campaign to rid the world of DINP and other compounds that have been proven safe in plastics. Dr. Moore describes "a campaign of fear to implement [the environmental lobby's] political agenda." He also said, "This fear campaign merely distracts the public from real environmental threats."

The campaign has been successful, as demonstrated by many retailers switching to phthalate-free products to avoid public pressure. Dr. Moore quotes the Consumer Product Safety Commission as saying: "If DINP is to be replaced in children's products . . . the potential risks of substitutes must be considered. Weaker or more brittle plastics might break and result in a choking hazard. Other plasticizers might not be as well studied as DINP." Despite the unknown risks of replacing DINP, retailers bowed to pressure from the environmental lobby and forced many products to replace DINP. As a result, we've removed a compound that has been proven safe through a multitude of testing around the globe and replaced it with untested compounds that may pose greater hazards.

Not only have large retailers yielded to the political power of the environmental lobby, entire countries and the European Union have also banned the use of phthalates in toys. In Europe, the ban on phthalates was passed before a government-sponsored evaluation of DINP was completed. The evaluation subsequently found no measurable risks tied to DINP, but the ban remained in place. Israel was among the first countries to ban phthalates, but they have since realized the dangers of non-tested substitutes. In Dr. Moore's words, "Israel realized the error of putting politics before science, and reinstated DINP."

Since Dr. Moore's editorial, environmentalists have found a new bogeyman in plastics: Bisphenol-A (BPA). Most of us are aware of recent studies indicating that BPA is a hormone disruptor and humans can absorb this substance from contact with certain plastics. In large enough doses, research indicates BPA can lead to developmental problems in children and a host of other health issues. That's generally all we need to hear in order to demand protection from BPA, but what if those *large enough doses* are so high, it's impractical for humans to actually be harmed? It's an issue we seldom consider and a question the media never asks. Understandably, we don't want anything in the environment that could harm our health, but let's not forget that drinking water in *large enough doses* can be fatal—and we don't want to rid the world of drinking water. Perhaps we're safer with BPA-free products, but are we certain the BPA substitutes are as safe and as well tested?

Like the Wakefield study on autism, you've probably heard a great deal about the dangers of BPA, but it's also likely that you haven't heard about numerous studies showing no health risks tied to BPA. Examples of such studies include the European Food Safety Authority (2010), a joint UN / WHO panel on BPA (2010), and the Advisory Committee of the German Society of Toxicology (2011). Also, on March 30, 2012, the FDA refused to ban BPA from food containers, indicating there were flaws in the scientific experiments identifying potential health risks. The FDA stated that the animal studies used to confirm health risks could not be translated to humans.

Environmentalists and the media immediately attacked this FDA decision. One of the more colorful media articles attacking this decision was titled "FDA Keeps Toxic Plastic Chemical in Food, Infant Formula." The title speaks for itself. In typical fashion, the media had no room for the mundane fact that the FDA found no risks tied to BPA. Instead, the media served as judge, jury, and executioner for both the FDA and BPA. Such media coverage is all too common, which means you and I generally hear only the sensational side of the story, but never the boring, unbiased facts.

Are the BPA warnings justified, or are they overly hyped fears equivalent to fearing the carcinogens in broccoli?

You can decide for yourself, but the real take-away from this BPA story is the level of influence the media and the environmental lobby have on our beliefs. That influence was on display when the FDA finally agreed to ban BPA in baby bottles and children's Sippy-cups. This ban, however, was not based on scientific findings of health risks. It was based on a request from the plastics industry, which had already replaced BPA in baby bottles and Sippy-cups due to the public's perception of BPA. The industry petition was issued to ease consumer fears—not because science demonstrated legitimate health risks.[71]

We have plenty of reasons to oppose plastics. Plastics are slow to degrade, they fill our landfills, and suspend in our oceans. As of this writing, bits of plastic from the 2011 Tsunami in Japan are washing ashore on Hawaii and are being found in fish and birds, often causing their death. There are more than enough reasons to oppose plastics. Phthalates and BPA offer two more reasons, but we can only hope the replacements for these ingredients are as well studied and as safe as the original compounds.

Dr. Moore concludes his article by saying, "We all have a responsibility to be environmental stewards. But that stewardship requires that science, not political agendas, drive our public policy."

I would add this: We shouldn't let a sensationalizing media dictate our beliefs or our public policies.

Golden Rice

Rice is a major source of nutrition for hundreds of millions around the world. It's an excellent source of calories, but it doesn't provide key nutrients that fight blindness and premature death in children. As a result, every year roughly half a million children go blind and 70% of those children ultimately die because rice is their primary source of nutrition.[72] Golden Rice refers to rice that has been genetically engineered to contain nutrients that prevent blindness and premature death.

Golden Rice is one of the new bogeymen for Greenpeace. According to Greenpeace, it's unnatural and therefore must be banned. Many agree with Greenpeace and strongly protest such intrusions on Mother Nature. Large agricultural firms are increasingly vilified for their role in developing genetically engineered crops. It's as though we believe genetically altered crops will unleash a ticking time bomb that will undo the very fabric

---

71. Sabrina Tavernise, "FDA Makes It Official: BPA Can't Be Used in Baby Bottles and Cups," *New York Times* (July 17, 2012), http://www.nytimes.com/2012/07/18/science/fda-bans-bpa-from-baby-bottles-and-sippy-cups.html (Accessed 12-8-12).
72. Dr. Henry I. Miller, "Save the Whales, Forget the Children," *Wall Street Journal* (October 31, 2012), A13.

of nature causing our ultimate doom. Are such fears justified, or have we simply seen too many science-fiction movies depicting such environmental cataclysms?

You'll have your own answer to that question, but consider this: some 85% of all corn and 91% of all soy products grown in the US are genetically modified and have been for nearly twenty years without causing harm.[73] If you're violently opposed to the use of such crops, are you certain you've been given all the facts? Genetically modified crops have greatly increased food production, reduced insecticide usage, reduced water usage, and allowed more "no-till" farming to reduce energy use and erosion. Is Gold Rice leading to our ultimate doom, or is it a miraculous, lifesaving breakthrough?

The attack on Golden Rice puts political agendas ahead of science and unfounded fears ahead of saving lives. It's not only irrational; it's inhumane. It's also typical of many irrational positions held by the environmental lobby—and subsequently held by many of us. The following examples will demonstrate.

The Mojave Ground Squirrel

Did you know the environmental lobby opposes solar energy in the Mojave Desert? That's because the Mojave ground squirrel is listed as one of California's endangered species and it lives—shockingly enough—in the Mojave Desert.

One project in particular is worth mentioning: A proposed 50-megawatt solar plant to be located just outside the Southern California International Airport (A megawatt is enough energy to light ten thousand, 100-watt light bulbs). It's believed that, at one time, the Mojave ground squirrel inhabited the site of this proposed solar plant. That alone was enough to merit environmental opposition, even though the ground squirrel no longer inhabits the area. This opposition drew the ire of Governor Arnold Schwarzenegger who was quoted as saying the project delay was caused by an endangered squirrel that "has never been seen on that land where they're supposed to build the solar plants. But if such a squirrel were around, this is the kind of area that it would

---

73. Ibid.

like, they say." [74] Apparently, its taboo to build anything that might one day interfere with a sunbathing ground squirrel.

Martha's Vineyard

The well-heeled residents of Martha's Vineyard fought construction of offshore wind turbines being proposed in their region. Numerous concerns were raised regarding the potential environmental impacts, but some believe the opposition was primarily driven by fear that the wind turbines would obstruct ocean views from million dollar balconies. These are purportedly some of the greenest people on the planet, but their environmental enthusiasm apparently stops within view of their mansions.

Save the Prairie Chickens

I was associated with a wind energy project proposed for a location in Kansas that would have produced some of the lowest-cost wind energy in the nation. Unfortunately, the location that made this project so economical was near the breeding grounds for Lesser Prairie Chickens. These prairie chickens are potential candidates for inclusion on the Endangered Species List and they may not reproduce in the presence of tall structures. Needless to say, 262-foot-high wind towers aren't a welcomed site for prairie chickens.

Strong environmental opposition forced the relocation of this project to a site with lower average wind speeds, meaning the project would produce less electricity and produce it at a higher cost than the preferred site. The move cost electric ratepayers a few million dollars over the life of the project, but we protected a potentially endangered species in the process.

While this was the proper decision, I found it interesting that the strongest environmental opposition came from sportsmen, a.k.a., bird hunters. The prairie chicken may be a candidate for Federal protection under the Endangered Species Act, but you can still buy a hunting permit for around $40.00 that will allow you to kill forty

---

74. Steve Williams, "Extortion and Environmentalism," *High Desert Daily Press* (May 23, 2008), http://www.vvdailypress.com/opinion/bureau-6570-environmentalism-extortion.html (Accessed 8-23-2009).

birds a season. We protected the prairie chicken at a cost of a few million dollars so bird hunters could kill a soon-to-be endangered species for the low price of one dollar per bird.

Wouldn't it make a lot more since to simply stop issuing hunting permits?

Save the Prairie Chickens, Again

New transmission lines have been proposed to allow wind projects to deliver energy from the US Midwest to more populated regions in the East. In order to protect one covey of prairie chickens, environmentalists, state and federal agencies, and bird hunters are forcing one of these new transmission lines to be rerouted.

This covey of prairie chickens is thought to include some 140 birds. Re-routing the transmission line is estimated to cost $567 million. You already know where this is going—electric ratepayers will pay over $4 million per bird to protect this covey so sportsmen can kill every bird in the covey for around $140.[75] As previously asked, wouldn't it make more sense to stop issuing hunting permits?

The Bear Essentials

Polar bears are so cute and adorable they're often used in commercials. The documentary film, *An Inconvenient Truth*, used cartoon polar bears to demonstrate the peril they faced from global warming. Due in part to public sentiment driven by this film and intense environmental lobbying, the polar bear was added to the Threatened Species List in 2008. For the record, environmentalists had hoped to list the polar bear under the *Endangered* Species Act, which carries stronger protections than being listed as a *Threatened* species. Although few would disagree with protecting the cute and adorable polar bear, the fundamental facts at the time challenged the need for any level of formal protection.

The global polar bear population was exploding when they were added to the list. Their numbers had grown from five thousand in the

75. "Costly power line shift would protect prairie chickens," Staff report, *Wichita Business Journal* (May 16, 2011),
http://www.bizjournals.com/wichita/morning_call/2011/05/costly-power-line-shift-would-protect.html (Accessed 1-22-13).

1960s to more than twenty-five thousand in 2008. Interestingly enough, regions that experienced a decrease in polar bear populations were areas where it has been getting colder over the past fifty years while growing populations occurred in areas where it has warmed.[76]

A report in the *Ottawa National Post* notes government surveys indicating that polar bear populations in one region had grown from 800 in the mid-1980s to an estimated 2,100 in 2007.[77]

Search the web for "polar bear populations" and you'll find numerous government sponsored surveys indicating this same population boom. You'll also find numerous articles and websites expressing the exact opposite and warning of the pending extinction of the polar bear. What are we to believe?

Most media stories depicting a decline in polar bear populations cite the Hudson Bay region. Between 1987 and 2004, polar bear populations in this region dropped from an estimated 1,200 to 950 polar bears: a decline of roughly 15 polar bears per year. What we're not told is that on average, 49 polar bears are legally shot and killed in this same region every year. Take away the hunters and the population would have grown. The media also fails to tell us that an earlier survey done in 1981 showed only 500 polar bears in the region. If we compare 1981 to 2004, we could argue that the population nearly doubled despite losses from hunting.[78]

It's difficult to know what to believe and who to trust in this matter, but consider the possible ulterior motives of the environmental lobby. By adding the polar bear to the Threatened Species List, environmentalists gained a powerful legal tool to stop future development in the Arctic. They can now pose legal roadblocks to slow or completely stop future attempts to access fossil fuel resources in that region. Is their victory good news or bad news?

---

76. Dr. Bjorn Lomborg, *Cool It*, 5.
77. Don Martin, "Polar Bear Numbers Up, But Rescue Continues," *Ottawa National Post* (March 6, 2007), http://www.nationalpost.com/news/story.html?id=1ea8233f-14da-4a44-b839-b71a9e5df868 (Accessed 8-11-09).
78. Dr. Bjorn Lomborg, *Cool It*, 6.

Globally, 300 to 500 polar bears are legally shot and killed every year.[79] Like the prairie chicken, if we were genuinely concerned about the fate of the polar bear, wouldn't it make a lot more sense to stop issuing hunting permits?

Save the Lizards

The environmental group, Wild Earth Guardians, has filed a petition to list the Spot-Tailed Earless Lizard as another endangered species. This lizard's home is part of the Eagle Ford Shale region of Texas: an area emerging as one the top oil and gas producing regions in the US. The Wild Earth Guardians are also anxious to classify the Dunes Sagebrush Lizard as endangered. This lizard happens to inhabit other regions that are prime areas for gas and oil production.

Here's the fun part of this story. Between 2007 and 2011, the Wild Earth Guardians collected nearly $700,000 from taxpayers though grants provided by the Fish and Wildlife Agency. They used the data generated from these grants to sue this same Fish and Wildlife Agency 76 times, alleging environmental violations.[80] That's more than one lawsuit every month for five years! Such lawsuits are a common tactic: bury an agency, a company, or an entire industry with frivolous lawsuits in hopes of stopping development or gaining concessions that significantly limit development. The tactic is so common it has a name: "Sue and Settle." How common is it?

Over the roughly 6,000 business days between 1990 and 2014, just two environmental organizations filed more than 1,000 lawsuits attempting to gain concessions from industry, slow development, or block access to reserves of fossil fuel.[81] That's roughly one lawsuit every week for twenty-four years—and it doesn't include lawsuits by the Sierra Club, Greenpeace, or a host of other green organizations.

---

79. Ibid.

80. David Porter, "Playing Chicken in Oil-Patch Politics," *Wall Street Journal* (December 6, 2012), A15.

81. "Sage Grouse Rebellion," *Wall Street Journal*, Review and Outlook, March 11, 2014, A14.

Save Everything

Environmentalists are hoping to add 757 new species to the list of protected wildlife by 2018. The primary species targeted for protection happen to inhabit regions that are also ripe for gas and oil development.[82] Although protecting species that are truly endangered is a noble cause, there seems to be a locational bias involved in the selection of threatened species. Are we certain that wildlife protection is the genuine motivation of environmental organizations?

Our efforts to save species are one way we're trying to make amends for our past bad behaviors. Our efforts, however, often result in questionable actions. One action we're now pursuing is to reintroduce animals to regions they once inhabited, but were either hunted-out or forced to move due to development. A prime example is the reintroduction of black bears into the wilds of the Missouri Ozark Mountains. Personally speaking, I don't relish the possibility of emerging from my tent one morning in a remote campground to find myself standing between a mama black bear and her cubs. One of the more surprising selling points for reintroducing black bears (and numerous other animals) is to eventually open hunting seasons for the creatures we're so determined to *protect*. This logic seems flawed.

An even more controversial move is California's recent decision to add the wolf to their list of endangered species. Wolves generally travel in packs that are little more than efficient killing machines. Although they tend to avoid humans, they won't hesitate to attack if we're the nearest source of dinner they can find. A small child would be an especially vulnerable target. I don't relish the idea of my granddaughter one day riding her bike through an area frequented by wolf packs. To protect humans and livestock from the dangers posed by wolves, communities around the world spent decades offering bounties for proof of wolf kills. Do we really want to encourage the resurgence of these well-equipped killers?

---

82. Ibid.

Hog Farms

I was involved in discussions with the Sierra Club to evaluate alternative ways to reduce emissions of greenhouse gases. One of the offers we proposed was to enclose hog farm waste lagoons and convert the waste into methane gas to generate renewable electricity.

It seemed like a win-win. The proposed projects would produce renewable energy, reduce greenhouse gas emissions, increase revenue for farmers, and reduce the threat of accidental spills that could contaminate groundwater. Energy produced from hog waste costs more than energy produced by many other renewable technologies and significantly more than energy from most existing fossil plants. Therefore, the only losers in this proposal were electric ratepayers who would have funded this higher cost.

At first, the Sierra Club's reaction was one of delight, but then it hit them: Supporting such projects would help pay farmers who run Confined Animal Feed Operations (CAFOs). The Sierra Club opposes CAFOs for a number of reasons, but primarily because they believe confining animals is cruel. Based on their strong objections, we withdrew our proposal.

In this case, the Sierra Club would rather stop CAFOs than 1) Protect ground water resources; 2) Reduce greenhouse gas emissions; 3) Provide renewable energy; and 4) Support the farmers who help feed a starving world. CAFOs will continue to operate with or without the offered waste lagoon clean up, but the Sierra Club couldn't support any project benefiting CAFOs—no matter how environmentally friendly the project might be.

Irrational Environmentalism

Opposing such things as solar energy in the Mojave Desert, "the biggest advance in the history of public health", and renewable energy from hog wastes isn't supporting the environment: It's opposing simple human progress.

When we stop programs that save millions of lives each year (DDT), we're pursuing irrational—and inhumane—environmentalism. When we fight against nutrient enriched Golden Rice, we're pursuing irrational—and inhumane—environmentalism. When we continue to issue hunting permits and choose instead to block access to low-cost energy reserves to protect a species that is currently thriving (polar bears), we're pursuing irrational environmentalism.

There's a reason we often react in irrational ways to environmental concerns. We've been trained from childhood to do exactly that.

Trained From Childhood

It's imperative that our children learn to respect and protect the environment. The question is whether or not they're learning the whole story.

In Steve Milloy's book, *Green Hell*, there's a section called "Compliance Starts Young." In Christopher C. Horner's book, *Red Hot Lies*, one chapter is called "Poisoning the Little Ones." Both books describe pamphlets and teaching guides that are compiled by environmental groups and distributed free to our schools. The books also discuss volunteers from local environmental groups who offer their time to educate our children about environmental topics.

While we appreciate the free educational assistance, we need to realize these lessons are not presenting an unbiased, balanced view. Our children are learning a one-sided view of environmental issues. Our children's teachers learned from these same lessons. Today's media journalists, and you and I have grown up learning this same one-sided view of environmentalism. Yes, it's vital that our children learn to protect the environment, but it's just as vital they learn to consider all sides of any issue.

Consider how we'd react if these free lessons were being provided by a hate group, the gun lobby, or Big Oil. We'd never tolerate such propaganda in our schools, yet when the biased learning comes from the environmental lobby, it has our full support.

The training isn't limited to classrooms. An example is a website titled *Pollution: a Guide for Kids by Tiki the Penguin*. This website is obviously designed for children who recently mastered the art of reading. In large, colorful lettering at the top of this webpage are words like *Muck, Stink, and Poison*. Tiki the Penguin instructs children by saying that pollution is everywhere, it kills plants and animals, and it's making the climate change. Tiki the Penguin leaves no doubt. The science is complete: Pollution kills, pollution causes climate change, humans are responsible for it, and we have to stop it—"especially you kids."[83] That's a lot of pressure for our three to five year olds.

From the day we're old enough to watch cartoons, we're trained to equate industry and energy with sickness, destruction, stink, poison, and muck. No one reminds us that industry and low-cost energy are also responsible for improved standards of living, increased food production, improved human health, a higher quality of life, and increased longevity. Protecting the environment is humanity's greatest responsibility, but there are two sides to every issue. If we fail to learn both sides, the result is often irrational.

Do we truly want our children to learn from groups that oppose renewable energy from hog farm wastes, chlorinated drinking water, and solar energy in areas of the Mohave Desert where squirrels don't even live?

An Organized Few

It's been said that an organized few can control the disorganized many. When groups like the Wild Earth Guardians can use $700,000 in taxpayer funding to file more than one lawsuit per month for five years, are they protecting our best interests, or are they blocking projects that would benefit the disorganized many? Are the Wild Earth Guardians truly worried about the Dunes Sagebrush lizard, or is there real goal blocking access to low-cost energy? Should we even care about their motives as long as they're protecting lizards and other creatures of nature?

We'll each have our own answers. Those answers will depend on our backgrounds, the training we've received since childhood, and the facts we've learned from school, the media, and other sources. Are these sources giving us all the facts?

Our air is cleaner today than it's been in the past forty years. Why isn't anyone sharing this good environmental news? Indoor air quality is a greater threat to our health than outdoor air quality. Why aren't we warned about this danger? In areas where outdoor air is at its worst, Mobile and Area sources are the largest cause. Are we

---

83. *OneWorld.net*, "Pollution: a Guide for Kids by Tiki the Penguin," http://tiki.oneworld.net/pollution/pollution_home.html (Accessed 5-6-12).

certain our focus on industrial pollution is in our best interest?

Chapter 8
The Employment Prevention Agency

"It costs $1 billion more per factory for me to build, equip, and operate a semiconductor manufacturing facility in the United States [than to build and operate the same plant overseas]"
—Paul Otellini, CEO of Intel

It costs Intel one billion dollars more to create jobs in the US than it does to create those jobs overseas. Mr. Otellini doesn't blame US labor rates or American productivity. Although he mentions US corporate tax rates that are among the highest on the planet, he primarily blames oppressive US regulations that make it difficult and costly to do business in the US.

There's no question: Regulations are vital. We wouldn't have achieved the emission reductions previously discussed without regulation. We wouldn't have cleaner rivers and safer roads without regulation. We wouldn't have today's consumer and worker protections without regulation. Historically, regulation has been a very good thing, but the health benefits of new environmental regulations continue to shrink while the cost of implementation continues to escalate. Is it possible to have too much of a good thing?

The Regulatory Cliff

As we've seen, since 1970 we've reduced critical air emissions 68%. Indoor air quality is now routinely a greater health threat than outdoor air. In regions suffering our most polluted skies, the major sources of pollution are homes, small businesses, and vehicles—not industry. Korea demonstrates that jobs and a strong economy are as important, if not more important, for a healthy society than overly restrictive regulations. Despite these unpopular and unspoken realities, the EPA has recently unleashed an unprecedented level of new regulations targeting industry, energy, and ultimately jobs.

In the four short years from early 2009 through 2012, the new policy making regulations announced by the EPA threatened to surpass the entire history of EPA rulemaking![84] The pace and scope of new regulations was so intense, Congress asked the Congressional Research Service to compile a report addressing the potential impacts. The report neither defended nor supported the actions of the EPA, but cited 37 of the most significant and controversial rules being implemented. The report also indicated, "Both Democrats and Republicans in Congress have expressed concerns . . . and introduced legislation that would delay, limit or prevent EPA actions." [85] Some members of Congress threatened to withhold EPA funding if the agency continued to pursue the Regulatory Cliff.

The press, the environmental lobby, and perhaps a majority of our population hailed the EPA's newfound environmental activism. Should we?

Industry called this unprecedented flurry of regulations the EPA's "Train Wreck" or the "Regulatory Cliff." The new regulations also earned the EPA a nickname: "The Employment Prevention Agency." According to the economic consulting group, National Economic Research Associates (NERA), just seven of the new Regulatory Cliff rules are projected to reduce the US Gross Domestic Product (GDP) by $38 billion per year and reduce US

---

84. The American Legislative Exchange Council, "Economy Derailed" (April 2012), Executive Summary, v, http://www.alec.org/docs/Economy_Derailed_April_2012.pdf (Accessed 1-26-13).

85. Congressional Research Service (CRS), "EPA Regulations: Too Much, Too Little, or On Track?," December 12, 2013, https://www.fas.org/sgp/crs/misc/R41561.pdf (Accessed 1-31-14).

disposable income by $29 billion per year. [86] These same seven rules are projected to increase electricity costs $15 billion per year and threaten over 800,000 jobs per year for twenty years. These estimated costs and job losses are the total impact *after* including the additional green jobs required to meet the new regulations.[87]

That's clearly not the message we've been given. Instead, we're told these new regulations will not only improve our health, they'll also drive economic expansion and create jobs. We're also told that claims of lost jobs and increased consumer costs are nothing more than industry sabre rattling: an industry-backed attempt to scare voters and reduce support for much needed health protections. Which story can we trust? Do we trust the EPA, the media, and the environmental lobby or an organization of "research associates" we know nothing about? Before deciding, let's dig a little deeper into the protections provided by the EPA, the benefits they claim to provide, and the job creation claims that are tied to the Regulatory Cliff.

Small Boilers

One of the EPA's Regulatory Cliff regulations specifically targets small industrial boilers. There are roughly 183,000 of these small boilers located in businesses and factories across the nation. That may sound like a lot of boilers, but they represent a small source of overall emissions. For example, my local municipal electric supplier has four boilers that fall under this regulation. Only one of these boilers operates all year. The others only operate sporadically during summer months. It's a small source of pollution, but the new regulation will force the retirement of two of these boilers and increase the cost to operate the other two. Jobs will be lost, the cost of electricity will increase, and the environmental

---

86. Dr. David Harrison, Dr. Anne E. Smith, Scott Bloomberg, Andrew Foss, Andrew Locke, Sebastian Mankowski, Meredith McPhail, Reshma Patel, and Dr. Sugandha Tuladhar, "Economic Implications of Recent and Anticipated EPA Regulations Affecting the Electricity Sector" (October 2012), Executive Summary " 2. Economics Impacts," and tables ES-3 and ES-4, National Economic Research Associates (NERA).
87. Ibid, ES-5 and ES-8.

improvement will be unnoticeable. Across the nation, this new regulation will provide similarly small emission reductions while ensuring the loss of thousands of existing jobs.

An August 2010 study by the economic forecasting organization IHS Global Insights concluded that this lone regulation would put nearly 800,000 jobs at risk, reduce labor payrolls by $38 billion, and reduce US GDP by as much as $63 billion.[88] More industry sabre rattling?

Needless to say, the EPA tells a different story. The EPA said this regulation would provide $1 billion to $2.4 billion in annual economic benefits[89] and would prevent:

- 24 to 61 annual premature deaths,
- 17 annual cases of chronic bronchitis,
- 40 nonfatal heart attacks per year,
- 40 hospital and emergency room visits per year,
- 38 annual cases of acute bronchitis,
- 800 annual cases of respiratory symptoms [whatever that means?]
- 3,200 days when people miss work or school per year,
- 420 annual cases of aggravated asthma, and
- 19,000 days per year when people must restrict their activities.[90]

---

88. IHS/Global Insight, "The Economic Impact of Proposed EPA Boiler/Process Heater MACT Rule on Industrial, Commercial, and Institutional Boiler and Process Heater Operators" (August 2010), 10, http://www.cibo.org/pubs/boilermact_jobsstudy.pdf (Accessed 1-26-13).
89. Judy Sheahan, "New EPA Rules, Guidance Could Prove Costly to Local Governments, Stifle Urban Revitalization," The United States Conference of Mayors, US Mayor Newspaper (September 20, 2010), http://www.usmayors.org/usmayornewspaper/documents/09_20_10/pg1_E PA.asp (Accessed 5-26-13).
90. US EPA, "Fact Sheet: Final Air Toxics Standards For Industrial, Commercial and Institutional Boilers at Area Source Facilities," 2, http://www.epa.gov/airquality/combustion/docs/20110221aboilersfs.pdf (Accessed 1-23-13).

These bulleted figures are extremely precise numbers based on very imprecise data—and a lot of assumptions. Engineers refer to such findings as highly precise, grossly inaccurate numbers. Precision is easily achieved: We can divide two numbers and get a *precise* answer down to several hundred decimals places. However, if the two numbers we divide are both guesses; our precise answer won't have any meaningful level of accuracy.

Arriving at such precise numbers is a combination of statistics gone wild and assumptions on steroids. The last bulleted item shows us why. With a US population of 300 million, those 19,000 days of "unrestricted activities" will provide an extra 9.5 seconds per person per year! How exactly did the EPA manage to develop this statistic?

The original version of this small boiler rule also had an interesting twist. If the small boiler was fueled by coal, it would have emission limits that were lower than the limits set for boilers using oil or natural gas. How does allowing more pollution from gas or oil-fired boilers protect anyone's health?

Save the Fish

Many industrial processes draw cooling water from rivers and reservoirs. The cooling water doesn't mix with any other substances, so the only impact is warming the water before it's returned to the original source. Pumping the water into industrial facilities, however, often results in killing fish. For decades, industry has been allowed to fund fish hatcheries to restock more fish than they were killing. Under the EPA's Regulatory Cliff, this practice is no longer acceptable. Industry is being forced to spend billions of dollars to prevent fish kills. The good news is we'll kill fewer fish. The bad news is the cost of doing business in the US will increase, jobs will be lost, and we'll ultimately have fewer fish. What's the wisdom behind this change in philosophy?

Farm Dust

If you want to build just about anything of size, you'll first have to obtain a Ground Disturbance Permit. One of the goals of this permit is to control the dust (particulate emissions) created by construction. Even if you're building on a site next to gravel roads where the surrounding trees are normally covered with gravel dust, you'll be held responsible for the dust you create on your construction site as well as any road dust attributed to the increased traffic generated by your project. You'll have to water your dirt and the gravel road to ensure your dust causes no harm.

Have you heard claims that the EPA wants to regulate farm dust? The EPA has assured us they're not in business to control farm dust, but it's only a matter of time before they do exactly that. To achieve such control, they could either expand the use of Ground Disturbance Permits or they could simply follow an established rule that addresses Regional Haze.

Regional Haze

Regional Haze isn't a new Regulatory Cliff rule, but it's worth discussing. This regulation is designed to improve visibility in national parks by eliminating any source of pollution that might restrict visibility in those parks. This 1999 regulation calls for a 64-year transition period that will end with visibility levels that are— according to the promotional language for this regulation—the same level of visibility that existed before humans existed. During the early portion of this sixty-four-year transition period, the EPA addressed pollution from stationary sources, primarily industry. As we get closer to the end of this 64-year transition period, the EPA will begin to address smaller sources of pollution including cars and farm dust if those sources impact visibility in national parks.

It's a wonderful goal for national parks, but I've always had one question. Who at the EPA is old enough to know what the visibility was before humans existed?

Regional Haze Part II

Oklahoma submitted a statewide plan for meeting one of the interim visibility targets along the 64-year glide path required by Regional Haze. When the EPA altered the plan, Oklahoma sued claiming the agency did not have legal authority to dictate how the state met the required target as long as the state met the target. Although that has traditionally been the legal precedent, the court change course and ruled in favor of the EPA. Under the EPA's plan,

Oklahoma electric ratepayers will fund an additional $1.2 billion project even though—according to the EPA's own analysis—there would be no appreciable change in visibility.[91] What's the EPA's reason for requiring the extra $1.2 billion from Oklahoma citizens if nothing is gained?

The Chain Link Fence

A colleague tells the story of permitting a new natural gas-fired electric generating plant. The plant passed all the environmental hurdles except the computer simulation model of its NOx emissions. The computer model indicated that NOx might accumulate on a small plot of land adjacent to the generator during a few hours each year. The chain link fence surrounding the site was the cause of this potential accumulation. The utility owned the land where this accumulation might occur and no one would ever live on this land, but the plant would not be allowed to operate as long as the computer model indicated this problem.

To solve this improbable impasse, my colleague simply moved the problem section of fence to jog around the trouble spot—problem solved. The air permit subsequently moved forward without a hitch. The cost impact in this case was minimal, but the situation demonstrates the often-absurd protections provided by overly stringent regulations. It's difficult to see how moving the fence protected anyone's health.

Protection Levels

To protect us from harmful air pollution, the EPA establishes a Reference Concentration, which is the level of exposure they consider safe. For air pollution, this Reference Concentration is defined as "an estimate (with uncertainty spanning perhaps an order of magnitude) of a continuous inhalation exposure to the human population (including sensitive subgroups) that is likely to be *without* [italics added] an appreciable risk of deleterious non-cancer health

---

91. Robert Varela, "Court Upholds EPAs Haze Plan," *Public Power Daily* (August 2, 2013).

effects during a lifetime."[92] To put that in simpler terms, seventy-years of continuous exposure to the EPA's Reference Concentration is *not* likely to harm the health of our most vulnerable populations: the very young and the very old. To account for the uncertainty mentioned in the definition, the EPA understandably errs on the side of caution. The agency generally establishes Reference Concentrations an order of magnitude (ten times) *below* the values it believes *won't* cause harm after seventy-years of continuous exposure.

Some see this as adequate protection, some see it as not enough protection, and others see it as overkill—protections that are so ultraconservative, they border on the ridiculous. You can make your own determination, but one thing is clear: No one stays in the most vulnerable populations (the very young or the very old) for seventy years.

For cancer-causing agents, the Clean Air Act requires the EPA to set safety margins that "do not reduce lifetime excess cancer risks to the individual[s] most exposed . . . to less than one-in-one million."[93] In other words, the EPA targets a one-in-one million risk of cancer, which they define as the likelihood that "up to" one person out of one million would contract cancer if continuously exposed to the reference dose for seventy-years.[94] Bear in mind that the final safety margin will likely be set an order of magnitude lower to ensure adequate protection.

Why have any risk? Don't we all want a zero-in-one-million chance of getting cancer from pollution? Of course we all want that, but what are we willing to sacrifice to achieve this ideal world? Such protections come at a cost, and when those costs begin to slow the economy, destroy jobs, and lower standards of living, our health suffers.

---

92. US EPA, Glossary of Terms,
http://www.epa.gov/ttn/atw/nata2005/gloss1.html (Accessed 9-5-12).
93. US Federal Registry / Vol. 73, No. 198/Friday, October 10, 2008 / Proposed Rules (60434), Part III Environmental Protection Agency 40 CFR Part 63, http://www.epa.gov/ttnatw01/rrisk/fr10oc08.pdf (Accessed 5-26-13).
94. Ibid.

Are the EPA's protections prudent and necessary, or are they so ultraconservative they're counterproductive? Consider this: If the EPA regulated highway speeds and determined that a seventy mile per hour crash might kill one-in-one million people, they would set the interstate speed limit at seven miles per hour—an order of magnitude below the "safe" limit. Of course, we know that more than one-in-one million people die in seventy mile per hour crashes, so the limit would be far below seven miles an hour.

Perhaps the current level of protection is what we need from the EPA. On the other hand, how many of us would happily agree to drive seven miles per hour in order to protect our health?

Protection from Dioxins

Dioxins are a byproduct of numerous manufacturing processes, but they're also naturally emitted through combustion from forest fires and volcanoes. Therefore, natural background levels of dioxin contamination can be found everywhere. Dioxins have been linked to a host of health problems including human cancers and human reproductive disorders. The literature on dioxins is terrifying. Dioxins have been called the most toxic chemicals known to science. That much is certainly true for the rodents used to test the effects of dioxins. Whether or not dioxins are equally toxic to humans is debatable.

Perhaps the two most often cited cases of civilian dioxin exposure are an industrial explosion in 1976 outside Seveso, Italy, and the contamination of Times Beach, Missouri. With little effort you can find articles discussing the ominous results of these events. With a little more effort, you can also find studies and articles concluding those ominous results were statistically insignificant and our fear of dioxin is exaggerated and unnecessary.

One such article reviews comments from Dr. Vernon Houk of the CDC who was among those who recommended the evacuation and cleanup of Times Beach, Missouri, in 1982.[95] Soon after

---

95. Irvine Reed, "The Dioxin Un-scare—Where's the Media?" Accuracy in the Media, inc., *Wall Street Journal* (August 6, 1991), A16:3, http://science.halleyhosting.com/sci/soph/fruitvale/dioxinscare.html (Accessed 9-22-12).

cleanup began, Dr. Houk indicated that he would not worry about the levels of dioxin at Times Beach because new scientific studies showed that low doses of dioxin are not the health risk once envisioned. He also criticized the rodent tests used to identify health risks in humans. According to Dr. Houk, rats tested for the effects of dioxin were fed such high doses that it essentially guaranteed that cells and organs would be attacked and cancer would result. The doses given to test rats were so high that it's virtually impossible for humans to experience similar exposure. He also said these tests don't necessarily translate between species.[96]

Ignoring such findings and choosing to err on the side of caution, in 2002, EPA guidelines for the safe levels of exposure to dioxin were 100 to 1,000 times lower than the standards set by Canada, Japan, Nordic countries, the US Agency for Toxic Substances and Disease Registry, the European Commission, and the Joint UN Food and Agriculture Organization / World Health Organization (WHO) Expert Committee on Food Additives.[97]

In 2010, the EPA bureaucrats determined that the already ultraconservative dioxin limits weren't strong enough. They lowered the dioxin standard to a level 270 times lower than the 2002 standard.[98]

The need for lower standards has been strongly challenged because the old guidelines were performing remarkably well. Emissions of dioxins dropped 90% between 1987 and 2000.[99] Between 1970 and 2000, the US body burden of dioxins (the levels

---

96. Ibid.

97. *DioxinFacts.org*, "A Comparison of Dioxin Risk Characterizations" (May 2002), 4, http://www.dioxinfacts.org/dioxin_health/public_policy/dr.pdf (Accessed 5-3-11).

98. Judy Sheahan, et al.

99. *DioxinFacts.org*, "Trends in Dioxin Emissions and Exposure in the United States," http://www.dioxinfacts.org/sources_trends/trends_emissions.html (Accessed 4-13-13).

found in the human body) decreased seven-fold.[100] A study of dioxin exposure around a chemical plant in Michigan found that the level of dioxin contained in whole foods—the source of over 90% of all human exposures to dioxin—skyrocketed between 1940 and 1960, but since 1990, exposure levels have dropped to roughly the same level seen in 1910.[101]

As for the jobs-impact of this new protection, the US Conference of Mayors indicated the new dioxin limits would bring a complete halt to urban redevelopment efforts that have been vital to job creation in our most blighted inner-city areas.[102] The impact on job losses won't stop with urban redevelopment projects. Given the dire warnings and the one-sided media coverage we've all seen, our fear of dioxin is understandable, but is that fear justified?

Poison Versus Poisoned

Of all the agencies and countries noted above, the US EPA is the only one that assumes any level of exposure causes harm. All of the other agencies say that no harm occurs at low levels of exposure. The difference between these assumptions is the distinction between being exposed to a poison and actually being poisoned or harmed. For example, that daily low-dose aspirin many take to reduce the risk of a heart attack would be considered a toxic poison under the EPA's view of dioxin. In fact, most over the counter medications would be banned. We'd also have to ban drinking water because drinking water in *large enough doses* can cause death. We'd have to forget about enjoying our favorite adult beverage. Even broccoli would be banned due to the carcinogens it contains.

---

100. *ScienceDirect.com*, "Dioxin risks in perspective: past, present, and future," http://www.sciencedirect.com/science/article/pii/S0273230002000442 (Accessed 10-17-13).

101. Timothy Towey, Yvan Wenger, Peter Adriaens, Shu-Chi Chang, Elizabeth Hedgeman, Avery Demond, and Olivier Jolliet, "Combined Intake and Pharmacokinetic Model to Predict Serum TCDD Concentrations," University of Michigan Dioxin Exposure Study, Figure 1 and Figure 2, http://www.sph.umich.edu/dioxin/PDF/OsloPosters/intake-pbpk_final (Accessed 1-5-13).

102. Judy Sheahan, et al.

Regulations to protect our health are vital, but there's a point where those protections go beyond scientific objectivity and rational thought. They become a political ideology that may be based on good intentions, but the results often create undesirable consequences while failing to measurably protect our health. It is possible to have too much of a good thing.

Mercury and the Numbers Game

One of the Regulatory Cliff regulations targets mercury from coal-fired power plants. For obvious reasons, we hate mercury and praise this new regulation. Do the fundamental facts justify our praise?

Coal-fired power plants are the largest source of human-caused mercury emissions in the US. That sounds ominous, but what does it actually mean? The vast majority of all mercury emissions come from nature. US coal plants produce just 1% of all global mercury and only 40% of those emissions are in the form that can impact fish.[103] Sixty percent of the mercury deposited in the US originates from sources outside the country and some two-thirds of all US based mercury emissions fall outside the US.[104] These facts won't change our fear of mercury or our hatred for coal, but here's the important part: The threat posed by US coal-related mercury emissions is so low, even the EPA could not identify health benefits tied to reducing those emissions!

This minor inconvenience, however, didn't stop the EPA from claiming the mandated mercury reductions would avoid 17,000 premature deaths each year, create 9,000 new jobs, and benefit the economy by some $140 billion.[105] If it couldn't identify any health

---

103. Brian H. Bowen, Marty W. Irwin, "Basic Mercury Data & Coal FiredPower Plants," (March 2007), https://www.purdue.edu/discoverypark/energy/assets/pdfs/cctr/outreach/Basics2-Mercury-Mar07.pdf (Accessed 12-11-12).
104. *MercuryAnswers.org*, "Hg and Power Plants," http://www.mercuryanswers.org/plants.htm (Accessed 2-5-12).
105. Susan Dudley, "EPA's risks outweigh rewards for new mercury rule," *The Hill* (12-20-11), http://thehill.com/blogs/congress-blog/energy-a-environment/200539-epas-risks-outweigh-rewards-for-new-mercury-rule (Accessed 12-11-12).

benefits, how did the EPA determine this rule would avoid premature deaths? It was easy.

The EPA simply applied co-benefits that were unrelated to mercury. These co-benefits came from reductions in small particulate emissions. However, these same particulate reductions were already required by other Regulatory Cliff regulations. The EPA is double-counted benefits.

According to Professor Susan Dudley (Director, Regulatory Studies Center, George Washington University and past Administrator of the Office of Information and Regulatory Affairs in the US Office of Management and Budget), over 99% of the health benefits claimed by the EPA have nothing to do with mercury and are entirely based on reducing particulate emissions.[106] We should note that prior to the Regulatory Cliff regulations taking effect, US coal plants were responsible for less than 3% of all US particulate emissions.[107]

Professor Dudley indicates that the EPA's model of reducing premature deaths "assumes causality, where none can be explained."[108] She also states, "Contrary to the EPA's claim that the rule will provide particular benefits to children; the premature deaths . . . are modeled to accrue to people with an average age of 80 years, who would live weeks or months longer, if at all, as a result of the regulations."

Professor Dudley and many others also challenge the EPA's claim of creating jobs and providing $140 billion in economic benefits. While there's no doubt that jobs will be created to install and operate the equipment required to meet the new mercury regulation, the job impacts don't stop there. This new rule increases the cost of electricity. As we've seen, when the cost of energy goes up, the cost of doing business goes up, the cost of living increases,

106. Susan Dudley, "EPA's risks outweigh rewards for new mercury rule," *The Hill* publication (12-20-11), http://thehill.com/blogs/congress-blog/energy-a-environment/200539-epas-risks-outweigh-rewards-for-new-mercury-rule (Accessed 12-11-12).
107. US EPA, "National Emissions Inventory (NEI) Air Pollutant Emissions Trends Data," http://www.epa.gov/ttnchie1/trends/ (Accessed 1-14-14).
108. Susan Dudley et al.

consumer prices increase, and the ripple effect on the economy can ultimately harm our health. We'll repeat this often, but consumer spending drives 70% of the US economy. What happens to that portion of the economy when we spend more for energy, have less to spend on consumption, and the cost of consumer goods increases? According to Professor Dudley, the EPA never considered any negative economic impacts stemming from their new mercury rule.

It's also common to believe that mercury contamination is a growing health concern due to mercury accumulating in fish. The fundamental facts don't support this perception. When a sample of Atlantic Blue Hake preserved during the 1880s was compared to 66 similar fish caught in the 1970s, there was no change in the concentration of mercury.[109] When scientists compared samples of tuna caught in 1971 to tuna caught in the same location in 1998, they found no increase in mercury concentrations in the newer fish. This finding occurred despite increased human-related mercury emissions over the same period. Scientists conducting this study concluded that naturally occurring deep ocean volcanoes and vents were the primary cause of mercury accumulation in tuna.[110] When Alaska's Public Health Department tested hair samples from 550-year old Alaskan mummies, the median mercury content was over four and a half times higher than the median mercury content of

---

109. Richard T. Barber, Patrick J. Whaling, and Daniel M. Cohen (Marine laboratory, Duke University, Beaufort, North Carolina 28516 and Los Angeles Natural History Museum, Los Angeles, California 90007), "Mercury in Recent and Century-Old Deep-Sea Fish," Environmental Science and Technology, 1984 18 (7), 552-555, http://pubs.acs.org/doi/pdf/10.1021/es00125a014 (Accessed 8-30-12) Copyright 1984, American Chemical Society.
110. Anne M. L. Kraepiel, Klaus Keller, Henry B. Chin, Elizabeth G. Malcolm, and François M. M. Morel, "Sources and Variations of Mercury in Tuna," Environmental Science and Technology, (2003), vol. 37 (24), 5551-5558, Copyright 2003 The American Chemical Society, http://pubs.acs.org/doi/pdf/10.1021/es0340679 (Accessed 8-30-12).

present-day Alaskans (2002).[111]

Our fear of mercury in fish may actually be doing more harm than good. Fish is commonly called brain food for good reason. The selenium in fish is vital for proper brain function (including brain development in the fetus) and it helps the overall growth and development of the fetus and young children. Seventeen of the top 25 dietary sources of selenium come from seafood.[112] In June, 2014, the US Food and Drug Administration (FDA) issued a report advising pregnant women and young children to eat more fish. The report stated that fish provide key nutrients that aid both brain development and early childhood growth.[113] It also indicated that eating more fish during pregnancy can actually improve your child's IQ.

Additionally, fish provide essential enzymes that support numerous biological functions.[114] The American Heart Association indicates that the natural Omega-3 fatty acids found in fish are critical for reducing heart disease. According to Dr. Dariush Mozaffarian, Assistant Professor of medicine and epidemiology at Harvard Medical School and the Harvard School of Public Health,

---

111. Dr. John Middaugh, "Alaska Advisory," Department of Health and Human Services, Food and Drug Administration, Center for Food Safety and Applied Nutrition, Food Advisory Committee: methylmercury (July 24, 2002),
http://www.fda.gov/OHRMS/DOCKETS/ac/02/transcripts/3872t2.htm
(Accessed 11-9-11).
112. Ibid.
113 US Food and Drug Administration (FDA), "New Advice: Some Women and Young Children Should Eat More Fish," June 10, 2014, http://www.fda.gov/ForConsumers/ConsumerUpdates/ucm397443.htm (Accessed 6-12-14).

114. US National Oceanic and Atmospheric Administration, Fisheries Service Pacific Islands Regional Office (Cooperative Agreement No. NA09NMF4520176), the US Department of Energy's National Energy Technology Laboratory (Cooperative Agreement No. DE-FC26-08NT43291), and the Energy & Environmental Research Center's Center for Air Toxic Metals® at the University of North Dakota, "Selenium and Mercury, Fishing for Answers," http://www.undeerc.org/fish/pdfs/Selenium-Mercury.pdf (Accessed6-6-13).

and co-author of one of the most comprehensive studies on the impact of fish consumption on human health, "The real danger in this country, the real concern, is that we're not eating enough fish. That is very likely increasing our rates of death from heart disease."[115]

Is our fear of mercury causing more harm than good?

Worth More Dead than Alive?

Placing a dollar value on human life is repugnant to most people, but not government bureaucrats. To justify the cost of new regulations, various regulatory agencies apply a dollar value to human life. The Transportation Department recently assigned a $5 million dollar value to human life to justify stronger roof supports on cars. The Food and Drug administration assumed a value of $6 million to justify adding pictures of cancer patients to packs of cigarettes. The EPA leapfrogged the competition by using a value of $9.1 million to justify the costs of the Regulatory Cliff.[116] No one is better at playing the numbers game than the EPA—and it's costing us billions while providing questionable health benefits.

New Source Review (NSR)

New Source Review (NSR) is not a new Regulatory Cliff regulation, but it's another rule worth examining. This regulation is tied to the Clean Air Act and it's one of the environmental lobby's favorite rules. The intent of NSR is a good one. It forces electric power plants to install the latest state-of-the-art air pollution controls if and when the owners of those plants make large investments to upgrade or extend the life of an existing power plant. The problem is no one can agree on a definition for the terms *large investment*,

---

115. Sora Song, "The Dangers of Not Eating Tuna," *Time Health & Family* (January 4, 2008),
http://www.time.com/time/health/article/0,8599,1706623-2,00.html
(Accessed 10-23-12).
116. Binyamin Appelbaum, "As US Agencies Put More Value on a Life, Businesses Fret," *New York Times online* (February 16, 2011),
http://www.nytimes.com/2011/02/17/business/economy/17regulation.html
(Accessed 5-21-11).

upgrade, or *life extension.* As a result, this cherished rule has done more to harm the environment than protect it.

To explain why, it's important to note that most existing coal-fired electric generating plants are over 25-years old. Therefore, new turbine technologies can be installed to significantly improve the efficiency of these plants at little cost. Improved efficiency means more energy from less fuel and with less pollution. This is a win-win: electric ratepayers save money, more energy is produced from fewer resources, and pollution is reduced. Under terms of NSR, however, these turbine modifications may be considered an *upgrade,* which requires the installation of new pollution controls. Adding the cost of new pollution controls means the turbine modifications are no longer economically viable. It also means the Public Service Commissions regulating investor-owned electric utilities won't support these projects.

While there's no doubt the environment would be better served with both the turbine modifications and the new emission controls, NSR has blocked hundreds of potential efficiency improvements for decades. As a result, we've spent decades burning more fuel, creating more pollution, and producing higher cost energy than necessary. The rule is well intentioned, but its application has caused more harm than good.

Supporting NSR is irrational thinking on the part of environmentalists everywhere. We'd be better served to repeal NSR and replace it with a rule that could only be interpreted one way—a rule that might be as simple as requiring electric power plants to install the latest state-of-the-art air pollution controls every twenty-five years. Of course, even this simple rule would be difficult to interpret as demonstrated by the EPA's Frankenstein unit.

The Frankenstein Unit

Any new electric power plant built in the US has to be equipped with the Best Available (emission) Control Technology (BACT). That may sound simple enough, but under the EPA's Regulatory Cliff, the definition of Best Available is changing. A look at control technologies for removing sulfur dioxide (SO_2) will explain.

The Best Available technology for capturing SO_2 is a Wet Scrubber, which can achieve 99% removal of SO_2. The next best technology for SO_2 removal is a Dry Scrubber, which can achieve 95-98% removal. When we look at SO_2 alone, the Best Available Control Technology is easy to see. However, wet-scrubbers emit more water vapor than dry-scrubbers. Under the calculations used by the EPA, this translates into higher levels of fine particulate emissions. Therefore, wet-scrubbers are not the Best Available technology for particulate emissions.

Historically, the EPA has allowed either technology (Wet or Dry Scrubbers) to count as Best Available, but the EPA has recently changed this interpretation. The EPA now defines Best Available to mean that SO_2 removal must meet the levels achieved with a Wet Scrubber and the level of fine particulate emissions must meet levels achieved with a Dry Scrubber. While that may seem reasonable to some, the technology to achieve both results doesn't exist. It's difficult to understand how a technology that doesn't exist could be called the Best Available—but such trivial issues don't stop the EPA.

Industry refers to this new EPA interpretation as requiring the creation of a Frankenstein Unit, a comparison to the work of Dr. Frankenstein who thought he could create a super human by assembling the best parts from different dead people.

Taxation without Representation

On December 7, 2009, the EPA determined that greenhouse gases endanger the public. Once the EPA reached this finding, it automatically gained the authority to regulate greenhouse gas emissions. The date is important because this decision was made roughly six months after the Senate killed the latest attempt to pass a law allowing the regulation of CO_2 emissions.

While many hailed this decision, consider the economic ramifications. The EPA's decision increases the cost of all forms of energy from fossil fuels. It's equivalent to a tax on energy. This decision increases the cost of anything that is planted, harvested, mined, transported, processed, or manufactured. It increases the cost of food and the cost to heat and cool our homes, cook dinner, watch TV, charge cell phones, drive to work, use a computer, or buy practically any consumer product.

More importantly, consider the Constitutional ramifications. EPA bureaucrats—political appointees we don't vote into power— decided to restrict greenhouse gas emissions six months after Congress—our elected representatives—made it abundantly clear they didn't want these restrictions. Do we really want non-elected government personnel to have the power to ignore the wishes of Congress? Do we really want a regulatory agency with the power to effectively tax everything we buy in direct defiance to a recent vote of our elected representatives? If so, we're supporting taxation without representation.

Of course, the EPA doesn't see it that way. I attended an energy conference where one of the EPA's Regional Directors told the audience that the EPA was following the wishes of Congress as spelled out in the Clean Air Act. The Clean Air Act was last amended in 1990! EPA officials blatantly ignored a six-month-old vote of Congress, and with a straight face, told us they're pursuing the wishes of Congress by referring to a thirty-year-old law. Nothing stops the EPA, not even Congress.

No matter how much you or I might agree with the need to restrict emissions of greenhouse gases, this EPA action goes against the very fiber of US democracy. The EPA has lost its way. It has done an end-run around the Constitution to pursue its own political agenda. This action is nothing less than taxation without representation; yet many hail the EPA's actions.

Granted, we're all frustrated with our do-nothing Congress. Many want the President or regulatory agencies to act where Congress fails to act. That's great as long as they act in accordance with our particular point of view, but it's not so great when their actions don't match our beliefs. Don't we need some level of checks and balances on the power of regulatory agencies? I understand the frustration with a do-nothing Congress, but perhaps a do-nothing Congress manages to do no harm. They certainly do less harm than the EPA.

EPA and Endangerment

The EPA's own Inspector General criticized the endangerment finding related to greenhouse gases indicating the EPA failed to obtain an independent peer-review as required by law.[117] The EPA's own Science Advisory Board (SAB) also criticized the Regulatory Cliff regulation requiring coal-fired power plants to meet certain carbon emission limits. According to the SAB, the required emission limits can only be achieved with a developmental technology called Carbon Capture and Sequestration (CCS). The Clean Air Act requires that all technologies mandated by the EPA be proven technologies that have "adequately demonstrated" the ability to perform at commercial scale and at a reasonable cost. CCS technology has never been proven outside the laboratory and no full-scale commercial operations have "adequately demonstrated" this technology. The cost, functionality, and safety of this technology are unknowns.

Nothing stops the EPA, not its own scientists, not Congress, and certainly not US law.[118]

Secret Science

The chairman of the House Committee on Science, Space, and Technology, Representative Lamar Smith, indicates that over a two-year period beginning in mid-2010, his committee sent six formal letters to EPA leaders and top White House officials requesting copies of the scientific data supporting the health and economic benefits tied to the EPA's Regulatory Cliff regulations. The Committee is still waiting.[119] The non-elected bureaucrats at the EPA apparently believe they don't have to answer to our elected representatives. If they have sound science on their side, why not share their information and eliminate any hint of bias or deception? What is the EPA hiding and what are its leaders afraid we'll learn?

---

117. Tennille Tracy, "EPA Criticized Over Greenhouse-Gas Findings," *Wall Street Journal*, September 28, 2011, http://online.wsj.com/news/articles/SB10001424052970204138204576598761880128654 (Accessed 1-31-14).
118. The *Wall Street Journal*, Review and Outlook, "Political Science at the EPA," December 24, 2013, A10.
119. Rep. Lamar Smith, "The EPA's Game of Secret Science," *Wall Street Journal*, July 30, 2013, A15.

Secret Emails

The economic impact of the EPA Regulatory Cliff is so dire, the EPA employed secret email accounts to handle "sensitive" subject matters thought to deal with the costs, pitfalls, and questionable science related to the Regulatory Cliff. Former EPA Administrator Lisa Jackson resigned from the EPA soon after it was revealed that she used a private email account to allegedly shield delicate discussions from public scrutiny under the Freedom of Information Act (FOIA).[120] By all appearances, this was a major scandal worthy of front-page headlines. Although the *Washington Times* carried the story, most mainstream media outlets completely ignored it.

Not to bring politics into the discussion, but didn't President Obama enter office pledging to run a *transparent* Administration? What happened to that pledge?

The Clean Air Act

The Clean Air Act has been a wonderful tool for the environment, but it's also the primary reason the EPA has such broad powers. Under terms of the Clean Air Act, the EPA has the power to decide what it can regulate—regardless of the wishes of Congress or the findings of its own internal review boards. The Clean Air Act essentially removes any system of checks and balances on the power of the EPA. This won't be a popular sentiment, but like NSR, it's time we repeal the Clean Air Act and replace it with a law that not only protects the environment and our health, but also our jobs, our economy, and the rights of US voters and consumers—not to mention establishing a law that forces regulators to follow the intent of the US Constitution and the votes of Congress.

Green Jobs and Fuel Poverty

We're told that greenhouse gas restrictions and Regulatory Cliff regulations will create green jobs to bolster the economy. Recent

---

120. Stephen Dinan, "EPA says Jackson has an internal use email, compliance with FOIA raised," *The Washington Times* (November 20, 2012), http://www.washingtontimes.com/news/2012/nov/20/epa-says-jackson-has-an-internal-use-email-complia/ (Accessed 2-14-13).

events in Europe and California strongly challenge this claim.

Europe is far ahead of the US in terms of strict environmental regulation, carbon taxes, and tax subsidies for green energy. A March 2009 University study in Spain reviewed green job creation in that country. The study's results, which were presented to the US House of Representatives, indicated that each green job created in Spain resulted in the loss of 2.2 traditional jobs and 9 out of every 10 green jobs created were dependent on the continuation of taxpayer subsidies.[121] Of course, these numbers are more of those dreaded statistics we know we should challenge. Are these statistics valid or just more industry sabre rattling?

If you have any doubts about the negative job impacts of green energy, read "Europe's Green Energy Suicide."[122] According to this article, to meet their CO_2 reduction targets, the UK plans to surround its coast with wind turbines at a cost of nearly $9,000 per household. The equivalent amount of energy from fossil fuels would cost less than $450 per household. The same article indicated that the UK has now coined the term "fuel poverty" to classify households forced to spend more than 10% of their income on energy. According to the article, some 18% of households now fit that category, but by 2016, the number will grow to nearly 43%.

We can look closer to home to see the job impacts of green regulation in California. Another article, "The Price of Green Virtue," discusses the impact of California's aggressive environmental laws including California Assembly Bill 32 (AB 32), which addresses greenhouse gases. As the article states, California's tough new regulations will lower the State's GDP by as much as 8.9% by 2020 and will force each household to spend an additional $2,500 per year for energy alone.[123] This figure doesn't include the increased

---

121. Testimony of Gabriel Calzada Alvarez, US House of Representatives, "Testimony before the House Select Committee on Energy Independence and Global Warming" (September 24, 2009), http://globalwarming.markey.house.gov/files/HRG/092409Solar/calzada.pd f (Accessed 10-12-12).
122. Rael Jean Isacc, "Europe's Green Energy Suicide," the *Wall Street Journal*, June 5, 2012, A15.
123. The *Wall Street Journal*, "The Price of Green Virtue The bill starts to come due for California's climate change law," July 9, 2012.

cost of any consumer product that required energy to manufacture, harvest or transport within the state's boundaries.

The most revealing item in the article tells us that 36% of the emission reductions created by AB32 will be achieved as the direct result of a slowing economy. In other words, the environmental improvement will come from lost jobs.

California's increasing energy costs and untenable regulations are already forcing many jobs to leave the state. Between January 2001 and May 2007 (prior to the Great Recession), California lost 21% of its manufacturing job base, or 400,000 jobs.[124] This occurred at a time when the rest of the nation was enjoying robust job growth. By December 2012, the losses climbed to 33% of the state's industrial job base, or 618,000 manufacturing jobs.[125] Over this same period, more than one million total non-farming jobs were lost, representing over 7% of total 2007 statewide employment.[126] By comparison, neighboring states (Arizona, Nevada, and Oregon) lost roughly 1% of their total 2007 employment.[127] The Great Recession no doubt contributed to the post-2007 job losses, but the anti-business, green regulatory atmosphere in California is a primary reason that state lost 7% of its jobs while neighboring states lost only 1%.

Business executives continually rate California as having the worst business environment in the US. Andy Puzder, the CEO of Hardee's Restaurants, indicates that it takes six months to two

---

124. California EDD, Labor Market Information Department, California Manufacturers & Technology Association, Manufacturing Employment Data, "California Manufacturing is declining 11% more than US 2001-2012," www.cmta.net/page/mnfg-trends.php. (Accessed 5-13-13).
125. California State Assembly, Assembly Republican Council, California Jobs First, "California's Job Problems, Facts and Figures," http://republican.assembly.ca.gov/caJobs/?p=problem (Accessed 5-13-13).
126. US Department of Labor, Bureau of Labor Statistics, http://www.bls.gov/eag/eag.ca.htm#eag_ca.f.6 (Accessed 1-8-13).
127. US Department of Labor, Bureau of Labor Statistics, http://data.bls.gov/timeseries/sms32000000000000001, http://data.bls.gov/timeseries/sms41000000000000001, and http://data.bls.gov/timeseries/sms04000000000000001 (Accessed 5-13-13).

years to secure permits for a new restaurant in California, but only six weeks to accomplish the same in Texas.[128] According to job relocation expert, Joe Vrancih, California was losing an average of 3.9 to 4.7 businesses a week to other states in 2010 and 2011.[129] These businesses didn't fail. They relocated to other states to avoid oppressive regulations, and they took their jobs with them.

These statistics aren't industry sabre rattling. They're the real-world impacts of strict regulation and a green economy. If The EPA remains unchecked to implement its Regulatory Cliff, the entire US economy will soon be paying the same jobs-price as the California economy.

We all want to protect the environment and save the planet, but are we certain the Regulatory Cliff is the right answer? Are we addressing our real environmental problems, or are we attacking perceived threats that will ultimately produce few benefits at great cost? One thing is certain: We won't get the fundamental answer to that question from the EPA, the environmental lobby, or the mainstream media.

Summary

It costs one billion dollars more to build and operate a microchip plant in the US rather than building it overseas. We blame corporate greed and Big Business for stealing our jobs, but we're blaming the wrong parties. A broken tax system is one problem, but an all-powerful EPA is a major reason the US has lost jobs to other countries. A lawsuit-happy environmental lobby is another contributing factor.

When a regulatory agency staffed with political appointees can snub the wishes of Congress, it's time we changed our view of that

---

128. Wendell Cox and Steven Malanga, "California — toxic for business," *Los Angeles Times* (November 14, 2011), http://articles.latimes.com/2011/nov/14/opinion/la-oe-cox-malanga-california-business-20111114 (Accessed 1-7-13).
129. John Fund, "California Dreamin'—of Jobs in Texas," *Wall Street Journal* (April 22, 2011), http://online.wsj.com/article/SB10001424052748704570704576275051374 356340.html (Accessed 1-7-13).

regulatory agency. When unelected bureaucrats dictate our future energy policies using private emails and secret science to shield their decisions, it's time we demand a change. When unelected bureaucrats can enact regulations that effectively tax every consumer in the nation—in direct opposition to US law and Congress—perhaps it's time we dismantled that all-powerful agency and start over.

Part III
The Immaculate Deception

"Everybody talks about the weather,
but nobody does anything about it."
—Mark Twain

Chapter 9
What We Know and Might Not Know About Global Warming

"Scientists . . . say they have confirmed beyond doubt that
climate change is being caused by human activity."
—February 18, 2005 ABC News Online[130]

If you've lived in the same region as long as I have, fifty-plus years, you know that global warming is real because you've experienced the warming firsthand. Living in the geographic center of the Continental US, I remember a childhood of snowy Thanksgivings with snow generally remaining on the ground from the first snowfall until the official arrival of spring. There were enough snowy months to encourage snowmobile dealerships to open in my hometown. By the time I was in college, the snows didn't come until late December and seldom remained on the ground more than a few weeks at a time—the snowmobile dealerships were gone. Some fifteen years later and well into my career, I was enjoying an annual New Year's Day round of golf on local courses.

The first thing we know about global warming is that it's very real. No one disputes the fact that our planet is getting hotter—at least, no one worth discussing. The debate, if you can call it a debate, essentially deals with two issues. The first issue is whether or not today's warming is abnormal. The second issue is whether or not humans are the primary cause. We could add a third issue by questioning whether or not the science is complete, but familiar quotes like the one at the top of this chapter have told us for more

130. "Scientists Say Global Warming is Undeniable," *ABC News* (February 18, 2005), The Polar Radar (for) Ice Sheet Measurements (PRISM), Global Climate Change Archive (February, 2005), http://www.ku-prism.org/polarnews/2005_glob_Environmental.htm (Accessed 11-5-13).

than twenty years that the science is settled. Is it?

What We Know about Global Warming

This will be redundant for most readers, but it provides a common background for later discussions. We know that 2000 to 2009 was the hottest decade on record, and by most accounts, the warming has continued through 2013. Global warming is very real.

If you saw the Al Gore interview on the November 3, 2009 *Late Night Show with David Letterman* or his later appearance in early 2013, you also know that human activities emit 90 million tons of CO_2 every day. Here are a few of the other facts we know about global warming:

- Earth has been warming at an alarming rate since the start of the Industrial Revolution—an event coinciding with society's rapidly growing use of fossil fuels.
- Burning fossil fuel emits CO_2, which is a heat trapping, greenhouse gas that acts to warm the atmosphere.
- Our use of fossil fuel is the primary reason we emit 90 million tons of CO_2 every day.
- Recent worldwide average temperatures have been "the highest ever recorded."
- Extreme catastrophic weather events have been occurring frequently around the world.
- As the direct result of recent warming, glaciers are retreating, polar ice caps are melting, and ocean levels are rising.

We know these facts with total certainty. In addition, we've been told that the overwhelming majority of scientists agree that humans cause global warming. We've also been told that scientists have found undeniable proof that humans are responsible for today's climate change.

These last two items stem from climate studies that began in earnest in 1988. That's when the World Meteorological Organization (WMO) and the United Nations (UN) Environment Program assembled a panel of scientists from around the world to study climate change. This panel of scientists is called the United Nations' Intergovernmental Panel on Climate Change (IPCC). The IPCC has issued five assessment reports summarizing the science behind global warming (1990, 1995, 2001, 2007, and 2013). Each subsequent report has identified increasingly convincing evidence that humans are responsible for global warming. In fact, the Summary for Policymakers from of the fifth IPCC assessment report (September 27, 2013) indicates, "It is extremely likely that human influence on climate caused more than half of the observed increase in global average surface temperature from 1951–2010." The IPCC indicates a 95% to 100% confidence level in this statement, which is an increase from 90% confidence in the IPCC's fourth assessment report.

Each IPCC report is a massive endeavor as demonstrated by the 450 lead authors, 800 contributing authors, and 2,500 expert reviewers who worked on the fourth assessment report (2007).[131] In case that's not enough evidence, new research indicates that 97% of all peer-reviewed climate studies agree with the IPCC. How could anyone debate the findings of so many experts and such overwhelming proof?

The Experts

Climatologists are generally acknowledged as the experts on Earth's climate, but it actually takes a wide variety of scientific disciplines to uncover the mysteries of Earth's climate system.

When it comes to the history of Earth, geologists are the original experts. They've been digging in the dirt to uncover Earth's history since long before global warming was an issue on anyone's mind. Geologists work side by side with archeologists, biologists, paleontologists, biochemists, botanists, astrophysicists, and a host of other experts in various disciplines to develop the history of Earth's past climates.

Scientists in these disciplines are the experts counting tree

---

131. Union of Concerned Scientists, "The IPCC: Who Are They and Why Do Their Climate Reports Matter?" http://www.ucsusa.org/global_warming/science_and_impacts/science/ipcc-backgrounder.html (Accessed 4-15-10).

rings to determine historic rates of growth and translating those growth rates into records of past climates. They're the ones analyzing the isotope ratios of marine sediment at the bottom of the Saragossa Sea to determine past levels of atmospheric CO_2. They're the ones looking at ice cores and the deep Earth to identify pollens, fossils, and other indicators to determine which plant and animal species thrived at different times in various regions of the world. These are the experts telling us that plants and animals have faced extinction and forced migration many times in the past as the direct result of climate change. They're the experts who give climatologists the data they need to do their work.

So, what work is left for climatologists? Climatologists take the data generated by these other disciplines and turn that data into predictions of future weather. To create these predictions, climatologists use their knowledge of meteorology and atmospheric science to develop computer simulation models designed to mimic Earth's climate system. They fill these computer models with the data provided by others and add their own set of knowledgeable assumptions. Subsequently, one of the primary roles for climatologists is to produce computer models that accurately duplicate Earth's historic climate behaviors and the climate's response to changing conditions. We should note that the IPCC's proof regarding humanity's impact on climate change rests almost entirely on the results of these computer simulation models.

Most descriptions of the college course work required for a degree in climatology mention the use and development of these computer models. The course descriptions also discuss classes that explore the various causes of climate change, "including the impact of human activities on climate change." This may seem inconsequential, but climatologists are trained to expect human involvement. Their training instills a preconceived ideology that can bias their research and the assumptions climatologists use in their computer models. Is this a minor consideration, or has it created a systemic, *groupthink* mentality that unknowingly prejudices the work of climatologists?

Computer Models and Proof

Computer models are invaluable tools for analyzing

tremendous volumes of data. They're used in a variety of fields and have proven infinitely helpful in many applications. They're helpful, but can they actually *prove* anything?

In simple terms, we can think of computer simulation models as strings of mathematic formulas tied together in an effort to duplicate real-world events. Perhaps the simplest example would be a computer model designed to mimic the flip of a coin. In this case, we know there are only two possible results—heads or tails. We also know the exact probability of getting heads or tails every time we flip the coin. No assumptions are required to precisely model the potential outcomes from flipping a coin. With such perfect knowledge, however, our model can't accurately predict the results of our next coin flip. Our model is no more accurate than guessing.

For a slightly more complex example, consider traffic planners and the number of assumptions they have to make in order to model traffic flow on a roadway for the next twenty years. How will demographics change in the area? Will gas prices lead to more carpooling, and if so, how much more? Will more people be working from home? Where will new apartments or neighborhoods emerge? Where will new jobs locate? Traffic planners have a tough job.

It's easy to see that as the number of variables and unknowns increase, the complexity of a computer model rapidly increases. In addition, as the number of unknowns increase, the accuracy of the model's predictions rapidly deteriorates. This mathematically proven conclusion can be found in a book titled *Beyond The Warming, The Hazards of Climate Prediction in the Age of Chaos*,[132] by Antony Milne.

If future traffic patterns are difficult to model accurately, consider the number of variables—and more importantly, the number of unknowns and uncertainties—involved in forecasting future climate. Climate models have to account for a wide range of constantly changing variables, including but not limited to, wind patterns, ocean currents, rainfall, humidity levels, cloud cover, snowfall, population growth, vegetation levels, solar irradiance, Earth's orbital patterns, cosmic rays, atmospheric aerosols,

---

132. Antony Milne, *Beyond the Warming: The Hazards of Climate Prediction in the Age of Chaos*, *Prism Press*, Great Britain, 1996.

reflectivity of ground cover, emission levels of some forty different greenhouse gases, and a host of other factors that can't be predicted with any meaningful level of certainty. Anthony Milne's book concludes that climate models are rendered useless by the vast number of uncertainties involved. He also advised that with enough repetitious modeling efforts, any observed future event— any change in climate—will be consistent with the model's results. We've all heard claims that recent climate events are consistent with the IPCC's computer simulation models. Simple math tells us why that's true and why it's a rather meaningless claim that doesn't validate IPCC model results.

The complexity involved in modeling Earth's climate is stunning, the number of uncertainties is staggering, and the level of required assumptions is practically unlimited. Is the science complete? Has the IPCC truly found "undeniable proof?"

The Debate

Despite all we've been told, the global warming debate continues among scientists with two schools of thought. Let's call these two schools of thought the Believers and the Skeptics. Believers agree with Al Gore and the IPCC: humans cause global warming. Skeptics believe natural climate drivers other than human CO_2 emissions are the primary cause of today's warming. It's likely most of us picked sides in this debate long ago. Most of us are also adamant in our beliefs, meaning we're not likely to change our opinions. Fortunately, we don't need to change our beliefs to realize there's a more important message contained within the global warming debate. That message focuses on the information we've been given, the groups and organizations providing the information, and most importantly, the information we haven't been given. The big story behind global warming is not choosing sides in the debate. The big story and the issue that should frighten us all is the way human-caused global warming has been presented to society. We haven't been given the whole story.

What We Might Not Know About Global Warming

Imagine how our view of global warming might have changed if the mainstream media had spent the last twenty-plus years highlighting the fundamental facts bulleted below. We'll explore each of these facts in more detail in later chapters:

- Historically, temperatures have always changed before levels of atmospheric CO_2 changed—it's never worked the other way around.
- Today's level of atmospheric CO_2 is quite low by long-term historic standards. CO_2 concentrations have been up to twenty-five times higher than today's levels without causing unstoppable global warming. Those high levels of CO_2 also failed to prevent ice ages from subsequently occurring.
- In their first assessment report, the IPCC shows graphs indicating that today's temperatures are above the average of the last 1,000 years, but below the average of the last 10,000 years.[133]
- This same IPCC report discusses two recent periods in history that were warmer than today, yet CO_2 levels were lower than today.[134]
- While humans emit 90 million tons of CO_2 every day, Mother Nature is emitting over 1.8 billion tons: more than 95% of all CO_2 emissions.
- When asked about past climate changes, the IPCC indicated that over the majority of the last 500 million years, Earth was much warmer than today and most

---

133. J.T. Houghton, G.J. Jenkins, and J.J. Ephraurms (eds.) "Report prepared for the Intergovernmental Panel on Climate Change by Working Group I," Cambridge University Press, Cambridge, Great Britain, New York, NY, UNIPCC first assessment report, 410, "Climate Change: The IPCC Scientific Assessment (1990)"; Chapter 7, "Observed Climate Variations and Change," Figure 7.1, 202, USA, Melbourne, Australia, 202. 134. Ibid, 202.

likely completely free of glaciers and polar ice.[135] How high were ocean levels back then?

- Because our planet still has permanent ice sheets and glaciers, scientists agree that Earth is currently in an ice age that began at least one million years ago.[136] We happen to be lucky enough to be living in one of many short and highly beneficial warm periods of this ice age.

- "Recorded history" for average global temperatures began at the precise time historians tell us the Little Ice Age was ending. The IPCC agrees: "It is clear that the period of instrumental [temperature] record began during one of the cooler periods of the past millennium."[137] If we started recording temperature at the end of an ice age, shouldn't we expect today's temperatures to be the highest ever recorded?

- NASA satellites indicate that nearby planets are warming

---

135. Solomon, S., D. Qin, M. Manning, Z. Chen, M. Marquis, K.B. Averyt, M.Tignor and H.L. Miller (eds.) "IPCC, 2007: Climate Change 2007: The Physical Science Basis." Contribution of Working Group I to the Fourth Assessment Report of the Intergovernmental Panel on Climate Change, Cambridge University Press, Cambridge, United Kingdom and New York, NY, USA, Frequently Asked Question 6.1,
http://www.ipcc.ch/publications_and_data/ar4/wg1/en/faq-6-1.html
(Accessed 11-16-11).
136. Science Channel, Ecosystems, "Are we currently living in an ice age?" Curiosity.com http://curiosity.discovery.com/question/are-we-living-ice-age (Accessed 9-30-13).
137. J.T. Houghton, L.G. Meira Filho, B.A. Callander, N. Harris, A. Kattenberg, and K. Maskell. "Climate Change 1995 The Science of Climate Change, Contributions of Working Group I to the Second Assessment Report of the Intergovernmental Panel on Climate Change." Section 3.6.4 (Summary of Section 3.6), 179.

in similar fashion to Earth. [138, 139, 140, 141]

- The high court in the UK ruled that Al Gore's Nobel Prize winning documentary film, *An Inconvenient Truth*, could not be shown in UK classrooms unless teachers first identify the film as a political work, not a work of science. Teachers are also required to read a list of claims made in the film that are not supported by science.

Cherry-Picked Data?

The above bulleted points are either well-established history or pertinent facts that have received little attention from the media, environmentalists, and others. Skeptics are routinely criticized for cherry-picking the data they present. If you've never heard the facts noted above, are you certain it's the Skeptics that are hiding the fundamentals and cherry-picking the data they choose to report?

More Unspoken Facts

The IPCC is divided into three distinct working groups. Only one group studied the physical cause of global warming. In other words, roughly two-thirds of the IPCC scientists never studied,

---

138. Lome Gunter, "Bright Sun, Warm Earth. Coincidence?" *National Post* (March 12, 2007),
http://www.canada.com/nationalpost/story.html?id=551bfe58-882f-4889-ab76-5ce1e02dced7 (Accessed 7-7-10). Also,
http://epw.senate.gov/public/index.cfm?FuseAction=PressRoom.Blogs&ContentRecord_id=469DD8F9-802A-23AD-4459-CC5C23C24651 (Accessed 7-7-10).
139. Kate Ravilious, "Mars Melt Hints at Solar, Not Human, Cause for Warming, Scientist Says," *NationalGeographic.com* (February 28, 2007),
http://news.nationalgeographic.com/news/2007/02/070228-mars-warming.html (Accessed 5-7-11).
140. James M. Taylor, "Mars Is Warming, NASA Scientists Report," *Environment & Climate News* (November 1, 2005),
http://www.heartland.org/policybot/results/17977/Mars_Is_Warming_NASA_Scientists_Report.html (Accessed 7-16-11).
141. Jonathan Leake, "Climate Change Hits Mars," *Times Online* (April 29, 2007), http://www.timesonline.co.uk/tol/news/uk/article1720024.ece (Accessed 8-12-08).

evaluated, or reviewed the possible cause(s) of today's warming. Believers say this is a non-issue because the IPCC still represents a lot of scientists. While that's certainly true, it seems someone should have told us that the vast majority of those scientists never studied or attempted to prove the cause of today's warming.

Speaking of proof, have you read any of the IPCC's assessment reports? If you do, you'll soon realize that counting sheep is overrated as a sleep aid. Even pounding down one of those high-powered, caffeine-and-herb-laced energy drinks won't provide much help staying awake through the dry language. If you do manage to trudge through the reading, you'll find the lavish use of words such as *likely, may, suspect,* and *assume.* These are not the words used to describe undeniable proof.

In fact, if you do a word search for the words *prove* or *proof,* you'll find a major void. When these terms are used, they're used in phrases like " . . . the lack of a *proven* physical mechanism"[142] or "A *proven* set of model metrics . . . has yet to be developed."[143] If you search the fourth assessment report for the word *prove,* you'll also discover that this report made what appears to be a few hundred *improvements* over the first three reports. The fifth report claims to have added even more improvements and more certainty. This leads to another important question or two. How do you improve on science that has allegedly proven your point and why would you continue spending billions of dollars per year on research once you had already proven your case "beyond doubt?"

---

142. Solomon, S., D. Qin, M. Manning, R.B. Alley, T. Berntsen, N.L. Bindoff, Z. Chen, A. Chidthaisong, J.M. Gregory, G.C. Hegerl, M. Heimann, B. Hewitson, B.J. Hoskins, F. Joos, J. Jouzel, V. Kattsov, U. Lohmann, T. Matsuno, M. Molina, N. Nicholls, J. Overpeck, G. Raga, V. Ramaswamy, J. Ren, M. Rusticucci, R. Somerville, T.F. Stocker, P. Whetton, R.A. Wood and D. Wratt; "(2007) Technical Summary In Climate Change 2007: The Physical Science Basis," Contribution of Working Group I to the Fourth Assessment Report of the Intergovernmental Panel on Climate Change [Solomon, S., D. Qin, M. Manning, Z. Chen, M. Marquis, K.B. Averyt, M. Tignor, and H.L. Miller (eds.)]; Cambridge University Press, Cambridge, United Kingdom and New York, NY, USA., 31.
143. Ibid, 87.

The fact is none of the IPCC assessment reports claim to have found proof of any kind, much less proof beyond all doubt. In their own words, the IPCC scientists indicate their conclusions are based on "expert judgment," which is to say it's all based on opinion.[144] It's allegedly the opinion of unbiased experts, but like it or not, this falls far short of *proof*. Only the media, the environmental lobby, and a few politicians are bold enough to interpret the IPCC reports as proof.

For most of us, this becomes a non-issue when we hear that 97% of all peer-reviewed science agrees with the IPCC. Of course, this 97% figure is another one of those bikini-like statistics. How much do we know about what's hidden underneath this figure?

Perhaps the most recognized source for this finding comes from a 2013 survey by John Cook and others.[145] This survey evaluated the Abstracts (the one or two paragraph summaries) of nearly 12,000 climate studies completed from 1991 to 2011. Based on that evaluation, only one-third (4,000) expressed an opinion regarding the cause of global warming. Cook's survey found that 97% of these 4,000 studies agreed with the IPCC. When the authors of those 4,000 studies were asked to self-rate their conclusions, however, only 1,400 responded indicating their

---

144. Lisa Alexander (Australia), Simon Allen (Switzerland/New Zealand), Nathaniel L. Bindoff (Australia), et al., "Working Group I Contribution to the IPCC Fifth Assessment Report Climate Change 2013: The Physical Science Basis
Summary for Policymakers," (September 27, 2013), Introduction, 2 of 32, http://www.climatechange2013.org/images/uploads/WGIAR5-SPM_Approved27Sep2013.pdf (Accessed 10-20-13).
145. John Cook, et al., "Quantifying the consensus on anthropogenic global warming in the scientific literature," (January 18, 2013), Environmental Research Letters, http://iopscience.iop.org/1748-9326/8/2/024024/pdf/1748-9326_8_2_024024.pdf (Accessed 10-7-13).

agreement with the IPCC.[146] Once again, Cook's survey found that 97% of the 1,400 respondents endorsed human-caused warming. This 97% *consensus* may sound convincing in the popular media, but the Skeptics in the crowd look at these results differently. They contend that two-thirds of the surveyed peer-reviewed climate studies (67%, or 8,000 out of 12,000) failed to reach a conclusion regarding the cause of today's global warming. Many also argue that less than 12% (1,400 out of 12,000) felt strongly enough about their findings to confirm their agreement with the IPCC.

Another often quoted survey indicating a 97% consensus highlights a few other concerns. This survey ranked climate expertise based on the number of published climate articles written by a scientist. Many scientists in this survey had published over 50 articles and some had published as many as 900 articles.[147] Correct me if I'm wrong, but being a prolific writer doesn't equate to subject matter *expertise*. It simply means that the writer has learned how to profit from the more than 100 billion dollars the US alone has spent to fund climate studies. Also, some of the scientists in this survey participated in studies that both agreed and disagreed with human-caused global warming! How are we supposed to interpret the beliefs of these scientists? More importantly, we don't know how many of the 12,000 reports in the above survey by Cook were written by the same scientist. If one scientist wrote 900 papers, the Cook survey counted that as 900 different scientific studies, yet all 900 studies clearly represent the beliefs of only one scientist.

Read the article "Global Warming Alarmists Caught Doctoring '97-Percent Consensus' Claims" or search the web for "97% consensus debunked," and you'll see there's more than a little

---

146. Brendan DeMelle, "Consensus Confirmed: 97 Percent of Climate Papers Agree on Manmade Global Warming," *Huffington Post* (5-15-13), http://www.huffingtonpost.com/brendan-demelle/consensus-confirmed-97-of_b_3282447.html (Accessed 9-12-13).
147. William R. L. Anderegg a, 1, James W. Prall b, Jacob Harold c, and Stephen H. Schneider, "Expert credibility in climate change" Proceedings of the National Academy of Sciences, April 9, 2010, Citation Analysis, http://www.pnas.org/content/early/2010/06/04/1003187107.abstract, (Accessed 11-7-13).

disagreement regarding the legitimacy of this 97% figure. You'll also find numerous studies and articles—some from past IPCC scientists—that flatly reject the 97% consensus.

So where does that leave us? What are we to believe?

The Bottom Line

Continue to believe whichever side of the debate you choose, but believe this: We haven't been told the whole story.

Chapter 10
History Repeats Itself

"Formerly the earth produced all sorts of fruit, plants, and roots. But now almost nothing grows. . . . Frost and cold torment people. The good years are rare. . . . Only a few take care of the *miserables*."
—Olafur Einarsson (Iceland, 1573 - 1659)[148]

In northern regions today, the birds of spring arrive earlier and stay longer than they did just a few decades ago. Armadillos now roam farther north than in recent years. Both plants and animals are thriving in regions farther north and at higher elevations than a generation ago. We're told these are signs of abnormal and dangerous climate change. Are they?

The Vikings
The quote at the top of this chapter describes life near the end of the early Viking settlements on Greenland and Iceland. Most of us learned about these settlements during our early school years. If we remember those lessons, we'll remember that the Vikings first settled in Iceland around 800 AD and expanded to Greenland around 950 AD. These settlements occurred during a climate period known as the Medieval Climate *Optimum*. The warm climate of that era allowed the Vikings to raise pigs, sheep, and dairy cattle. They were able to plant and harvest grains for nutrition and hay for their animals. They collected wild berries and other natural foods to support a robust and healthy lifestyle.

All of that began to change sometime around 1300 when the Little Ice Age began. The Little Ice Age lasted from approximately

---

148. Bryson, Reid A. and Thomas J. Murray (1977), "Climates of Hunger," the University of Wisconsin Press, Madison.

1300 AD until roughly 1870. A slight warm up is thought to have occurred in the 1500s, with the coldest climate experienced between 1600 and 1800.[149, 150]

As the cold took over, growing seasons became too short to support grain production. It became impossible to feed livestock, so dairy cattle, pigs, and sheep could no longer be sustained. It's thought the Vikings had to abandon farming entirely and rely on the sea for all of their nutrition. Eventually, fishing became less productive as the fish moved farther south and ocean ice around the settlements made fishing hazardous. Conditions became so dire that the average height of the Vikings began to decline as each new generation suffered increasing malnutrition—much like today's North Koreans.[151]

A Comparison

Compare this long-ago climate to a 2009 article discussing Greenland's climate today:

> Known for its massive ice sheets, Greenland is feeling the effects of global warming as rising temperatures have expanded the island's growing season and crops are flourishing. *For the first time in hundreds of years, it has*

149. Environmental History Resources, "The Little Ice Age, Ca. 1300 – 1870,"
http://www2.sunysuffolk.edu/mandias/lia/decline_of_vikings_iceland.html (Accessed 12-12-12).
150. H.H. Lamb, "Climatic Fluctuations," H. Flohn (ed), "World Survey of Climatology. Vol.2." *General Climatology* (New York: Elsevier, 1969), 236; and Schneider, S. H., and C. Mass, "Volcanic dust, sunspots, and temperature trends," *Science*, 190 (1975) 741-746.
151. Scott A. Mandia, "The Decline of the Vikings on Iceland," Taken from "Climate, History and the Modern World,"
http://www2.sunysuffok.edu/mandias/lia/decline_of_vikings_iceland.html (Accessed 11-8-11).

become possible to raise cattle and start dairy farms[152]
[Emphasis added].

When we read articles indicating that various parts of the world are "feeling the effects of global warming," we interpret that as bad news: a sign of dangerous, abnormal climate change. Is it?

Far from being dangerous, we're blessed to live in a warm period when it's possible to grow food and raise livestock in vast areas of the world that were once—and very recently—too cold to sustain such activities. Perhaps there's a reason historians called Earth's most recent warm period a climate *Optimum*?

As far as being a sign of abnormal climate change, the 2009 news story couldn't be clearer: "For the first time in hundreds of years," Greenland is enjoying the same warmth enjoyed by the Vikings some 500 to 1,200 years ago—when CO_2 levels were lower than today's level. Something other than CO2 obviously caused this recent climate optimum.

The Little Ice Age changed where and how humans lived. Some of those changes are still with us today. The popularity of beer in North America is one of those lingering impacts.

Prior to the Little Ice Age, wine was the popular adult beverage across Europe, including England and Northern Europe. Wine, of course, is generally made from grapes, while beer is made from barley and hops. Barley and hops thrive in colder climates, but grapes require longer, warmer growing seasons. With the arrival of the Little Ice Age, grapes could no longer be grown in England or Northern Europe, so the inhabitants of those regions turned to barley and hops to make beer. Modern day adults in the US and Canada have traditionally preferred beer over wine because most early immigrants to these new lands came from Northern Europe— at a time when the Little Ice Age and its lingering effects prevented grape harvests. The Little Ice Age is thought to be one of the reasons so many early immigrants to North America came from

---

152. Gerald Trauffetter, "Arctic Harvest: Global Warming a Boon for Greenland's Farmers," *Spiegel Online International* (August 30, 2006), http://www.spiegel.de/international/spiegel/0,1518,434356,00.html (Accessed December 2, 2009).

Northern Europe—they were hoping to escape the cold. When they came, they brought their traditions, their agriculture, and their taste for beer to the New World.

The Medieval Climate Optimum and the Little Ice Age are relevant climate events that we'd be foolish to ignore, yet that's precisely what we've been told to do. Should we?

One graph from the IPCC's first climate assessment report (2009) tells us everything the media and others have refused to share. Figure 10-1 is my rough recreation of the IPCC's critical graph.[153]

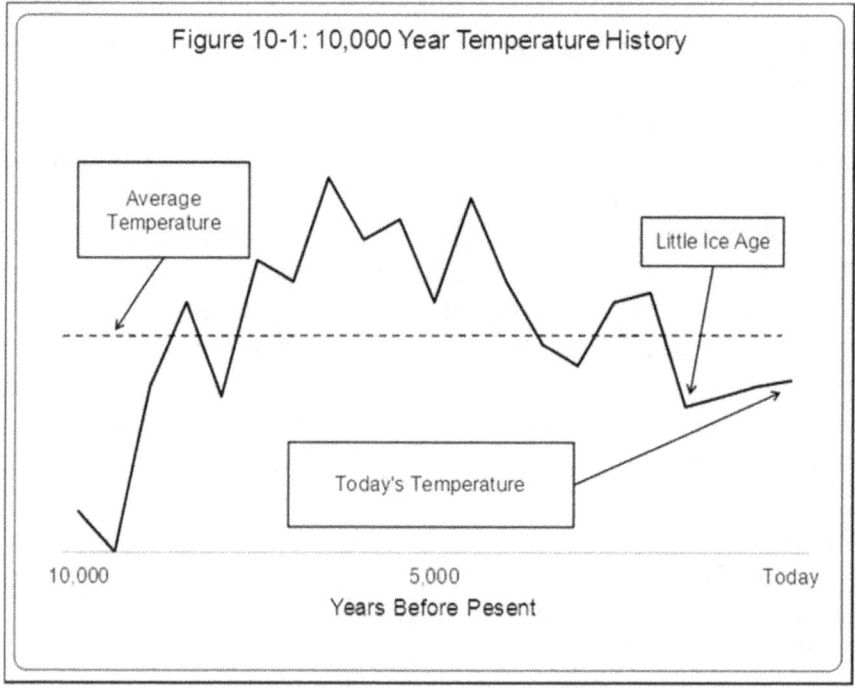

Figure 10-1: 10,000 Year Temperature History

Average Temperature

Little Ice Age

Today's Temperature

10,000   5,000   Today

Years Before Pesent

This graph begs a question or two. Why are we being told that today's temperatures are abnormally high and a frightening omen of pending disaster? Also, why should we believe that today's weather is more severe than it was during any past warm period in Earth's history?

153. J.T. Houghton, et al.

Our Glacial Past

Our early school years also taught us about past ice ages that lasted much longer and were far more severe than the Little Ice Age:

- Some 20,000 years ago Chicago, Illinois, was buried under a mile thick glacier.[154]
- At its maximum southerly progression, this glacier covered roughly 90 percent of Illinois.[155]
- Due to the extreme cold, the plants and animals that survived in the Midwest 16,000 years ago were vastly different from today's plants and animals.[156]

If we're worried about rising ocean levels, we should note that glaciers have been melting and refreezing and oceans have been rising and falling for the greater part of the last 20,000 years. The melting ice and rising oceans were interrupted and reversed by events like the Little Ice Age, but that mile thick glacier over Chicago didn't disappear overnight. Rising ocean levels are nothing new, nor are they a sign of abnormal climate change.

When we consider Earth's history of repetitive climate changes, today's climate is hardly anything new. Is it wise to allow the media and others to convince us otherwise?

---

154. Illinois Department of Natural Resources, "Landforms Tell the Story," http://dnr.state.il.us/lands/landmgt/parks/l&m/corridor/geo/geo.htm (Accessed December 9, 2009).

155. Blake Schneider, Leah Schwendeman, Rebecca Rice, Sarah Bourdo, Zach Roberg, "Geologic History of Illinois, "http://www.schools.lth5.k12.il.us/bths-east/geohist.html (Accessed 10-15-13).

156. Rickard S.Toomey, III; Erich Schroeder; Russel W.Graham; Eric C. Grimm; Pietra G. Mueller; Jeffrey J. Saunders; and Bonnie W. Styles (Sue Huitt, Editor, 2002, 2005). "The Midwestern United States 16,000 Years Ago," Illinois State Museum, R. Bruce McMillan, Director; http://www.museum.state.il.us/exhibits/larson/ (Accessed December 9, 2009).

Today Is Different

Believers provide many reasons to explain why we can ignore past climate changes. The most common claims I've encountered are shown below:

- Today's warming is occurring faster than any previous climate change.
- We don't know enough about past climate events to compare those events to today's climate change.
- The Medieval Climate Optimum and the Little Ice Age may have been regional events, not global events like today's warming.
- Today is different because humans are emitting 90 million tons of CO_2 every day.
- The overwhelming majority of scientists agree that today is different and humans are to blame.

Rapid Warming

Even the most devoted Skeptic would have to admit that if today's climate change is faster than any previous climate change, something is different. They'd also have to agree that human use of fossil fuels is the most obvious difference. Fortunately or unfortunately, depending on your point of view, the record of historic climate change is filled with large and abrupt climate changes as demonstrated below.

The National Research Council indicated that over the last 100,000 years, Earth has experienced a number of extreme climate shifts that occurred over a period of time of one decade or less.[157]

According to a report published in the National Academy of Science, "As the world slid into and out of the last ice age, the general cooling and warming trends were punctuated by abrupt

157. National Research Council (NRC), Committee on Abrupt Climate Change, 2002: "Abrupt Climate Change, Inevitable Surprises," *National Academy Press*, Washington DC.
http://www.nap.edu/catalog.php?record_id=10136 (Accessed 11-9-12).

changes. Climate shifts up to half as large as the entire difference between ice age and modern conditions occurred . . . in mere years to decades." This study also indicated that signs of abrupt past climate changes were found in Greenland, Canada, and South America: an indication that rapid climate change occurs on a global scale, not a regional scale.[158]

A study from the University of Michigan dealing with past climate changes indicated that "Evidence for this sudden cooling [10,000 years ago] . . . showed a [n] abrupt change from pre and postglacial forests to glacial shrubs and then back again. . . . The Younger Dryas provides dramatic evidence for rapid jumps in climate."[159]

A report from the Weather Underground said, "But what astonished them was the rapidity with which these [past climate] changes occurred. . . . These sudden climate changes affected not just Greenland, but the entire world. During the past 110,000 years, there have been at least 20 such abrupt climate changes. . . . 'Normal' climate for Earth is the climate of sudden extreme jumps— like a light switch flicking on and off."[160] This study also indicated, "The historical records show us that abrupt climate change is not only possible—it is the normal state of affairs."

The important points in the above discussion indicate that today's climate change is not unusual, it's not occurring more rapidly than past climate changes, and we can compare today's changes to those of the past. It's also apparent that past climate changes have impacted the entire planet, not just one hemisphere. Even the IPCC

---

158. Richard B. Alley, "Ice-core evidence of abrupt climate changes," Proceedings of the National Academy of Sciences, PNAS 2000 97 (4) 1331-1334; doi:10.1073/pnas.97.4.1331, http://www.pnas.org/content/97/4/1331.full (Accessed 2-1-13).

159. Regents of the University of Michigan, "Past Climates on Earth," Paleoclimates, Deep Sea Sediment Analysis, http://www.globalchange.umich.edu/globalchange1/current/lectures/kling/p aleoclimate/index.html (Accessed 11-15-12).

160. Jeffrey Masters, Ph.D. "The Science of Abrupt Climate Change: Should we be worried?" http://www.wunderground.com/resources/climate/abruptclimate.asp (Accessed 2-1-13).

agrees: "Some fluctuations [past climate changes] have nevertheless lasted several centuries, including the Little Ice Age which ended in the nineteenth century and was global in extent."[161]

What Caused Past Warm Periods?

Scientists routinely indicate that past climate changes were caused by a combination of changing volcanic activity, changing solar activity, and regular changes in Earth's orbit around the sun. Levels of greenhouse gases have also changed throughout Earth's history. The question is whether greenhouse gases or other natural events played the major role in causing past climate changes.

As previously noted, the IPCC indicated two recent warming events when temperatures were higher than today, yet CO_2 levels were lower than today.[162] In other words, some natural climate drivers other than CO_2 caused Earth's two most recent warm periods. Are we now supposed to believe those natural climate drivers stopped impacting Earth's climate because humans discovered fire?

There is one thing we know about the cause of long-ago warm periods. Human activities were not the cause.

150 Years of Headlines

Since we've been advised to ignore past climate change and focus entirely on the changes that started with the Industrial Revolution, let's do exactly that. Let's see what the last 150 years of climate change can tell us.

As mentioned in the first chapter, global cooling occurred from the mid-1940s to the mid-1970s. The 70s media focus was on the

---

161. J.T. Houghton, G.J. Jenkins, and J.J. Ephraums, eds., "Climate Change: The IPCC Scientific Assessment (1990)," Report prepared for Intergovernmental Panel on Climate Change by Working Group I (1990), Cambridge University Press, Cambridge, Great Britain, New York, NY, USA and Melbourne, Australia; C.K. Folland, T.R. Karl, K.YA. Vinnikov, Chapter 7, Observed Climate Variations and Change, 201. http://www.ipcc.ch/publications_and_data/publications_ipcc_first_assessm ent_1990_wg1.shtml (Accessed 11-16-11).
162. Ibid, 202.

coming ice age. Skeptics point to this media buzz as an indication of how little we know about climate, how quickly scientific beliefs change, and how quickly temperature trends change. On the other hand, Believers say this 1970s media buzz wasn't universally accepted like today's global warming. They say the overall upward trend over the last 150 years is the important issue. They say today's science is far more advanced with faster computers, better models, and improved knowledge. These statements are all true.[163]

Believers also say—interestingly enough—that the 1970s ice age fears were driven by a sensationalizing media that focused on only one side of the story.[164] Interesting theory!

Climate change has always been big news. We can thank research by Noel Sheppard, the Associate Editor of *News Busters*, for pointing out the following *New York Times* headlines covering roughly 150 years of climate history:[165]

- "The Hudson River, by a singular freak of temperature, has thrown off its icy mantle and opened its waters to navigation [This1870 headline coincides with the official end of the Little Ice Age]." —*New York Times* (NYT), January 2, 1870
- "The older inhabitants tell us that the winters are not as cold now as when they were young."—NYT, June 23, 1890
- "America is believed by Weather Bureau scientists to be on the verge of a change of climate, with a return to increasing rains and deeper snows and the colder

---

163. John Feeney Ph.D., Coby Beck, Eli Rabett Ph.D., Michael Tobis Ph.D., Andrew Dessler Ph.D., "A Wooden Stake in Newsweek's Global Cooling Heart," (November 4, 2006), *LogicalScience.com*, http://logicalscience.blogspot.com/2006/11/wooden-stake-in-newsweeks-global.html (Accessed 12-18-12).
164. John Russell, "What were climate scientists predicting in the 1970s?" (August 19, 2010) *Skeptical Science*, http://www.skepticalscience.com/ice-age-predictions-in-1970s.htm (Accessed 12-18-12).
165. Noel Sheppard, "150 Years of Global Warming and Cooling at the New York Times," (March 26, 2007), *NewsBusters.org*, http://newsbusters.org/node/11640 (Accessed 12-18-12).

winters of grandfather's day."—NYT, by the *Associated Press*, December 16, 1934

- "The theory that the world is growing slightly warmer is receiving added confirmation from temperature data."—NYT, February 15, 1959
- "After a week of discussions on the causes of climate change, an assembly of specialists from several continents seems to have reached unanimous agreement on only one point: it is getting colder."—NYT, Walter Sullivan, January 30, 1961
- "The earth, with few regional exceptions, is undergoing "a persistent cold wave" that began in the Nineteen Forties."—NYT, October 8, 1961
- "The United States and the Soviet Union are mounting large-scale investigations to determine why the Arctic climate is becoming more frigid, why parts of the Arctic sea ice have recently become ominously thicker and whether the extent of that ice cover contributes to the onset of ice ages."—NYT, Walter Sullivan, July 18, 1970
- "After invading Nebraska and Colorado, the armadillos, faced with increasingly frigid weather, are in retreat from those states toward their ancestral home south of the Mexican border."—NYT, January 27, 1972
- "From a study of ice extracted from deep within the Greenland ice sheet it appears that 89,500 years ago something catastrophic changed the climate from being warmer than today's to that of a full-fledged ice age."—NYT, Walter Sullivan, February 5, 1972
- "Scientists are reviving the controversial notion that millions of cubic miles of Antarctic ice can sometimes abruptly slip off the continent into the sea, resulting in extreme increases in global ocean levels and precipitating a dramatic chilling of the world's climate."—NYT, Walter Sullivan, March 9, 1980

The only thing normal about climate is that it's always changing. This fact is strongly reinforced by an article from the US National

Oceanic and Atmospheric Administration's (NOAA) archive that may sound familiar. Key excerpts from the article are paraphrased below:

> The Arctic is rapidly warming with unheard-of high temperatures. Water temperatures are some 12^0 Celsius warmer than historic averages. Ice is rapidly disappearing and is at record low levels. Seals and white fish have been forced to abandon their historic schooling grounds and move north to find colder conditions. Within a few years, it is predicted that due to the ice melt, ocean levels will rise and make most coastal cities uninhabitable.[166]

These dire warnings are based on a 1922 report regarding conditions in the Arctic. According to *Snopes.com*, the original *Associated Press* report was carried by the *Washington Post* on November 2, 1922, page 2. Of course, these dire warnings were later reversed during the climate chill our planet experienced from the 1940s to the mid-1970s.

This points to a fact the media and environmentalists routinely fail to discuss: Despite lower levels of CO_2, Earth warmed faster prior to the 1930s than it's warming today. What drove that warm period and why has the warming slowed despite increased atmospheric concentrations of CO_2?

In 1922, we didn't have satellites tracking Arctic sea ice and our surveys of the extent of ice were rudimentary and incomplete at best. When we're told that today's Arctic sea ice is at all-time lows, we have to realize that "all-time lows" means roughly the lowest levels over the last forty to fifty years! It's complete deception to believe that today's ice levels are unprecedented—especially when we know the IPCC indicated that today's temperatures are below the average of the last 10,000 years.

Simply put, Earth's climate has changed throughout millennia, and it has changed repeatedly over the last 150 years. Are the

166. George Nicholas Ifft, "The Changing Arctic," October 10, 1922, American consul [sic] at Bergen, Norway, published by *NOAA*, *Monthly Weather Review*, November, 1992, 589, http://docs.lib.noaa.gov/rescue/mwr/050/mwr-050-11-0589a.pdf (Accessed 3-12-14).

changes over? Is warming the only change we'll see over the coming decades and centuries? Perhaps today's severe weather will provide a clue.

Chapter 11
Weather Proof

"[Al] Gore points out the increase in wildfires, the melting glaciers, and gradual drying up of all continents as proof of global warming."
—Good Morning America[167]

We don't need declarations like the one above to know that global warming is real. The evidence is everywhere we look. A 2010 *Associated Press* news article serves as an example.

The article, "Climate breakdown? U.N. chief less optimistic now," discussed 2010 climate cataclysms, including record-breaking heat and drought in Moscow; record floods in China, Pakistan, and the US; and a 100-square-mile iceberg that fell from a glacier in Northwest Greenland.[168] The article indicated that the IPCC has warned for years that such disasters were imminent. According to the article, the 2010 climate disasters fit perfectly with IPCC predictions. The article also warned, "It's not just a portent of things to come, scientists say, but a sign of troubling climate change already underway."

The severe weather hasn't slowed since that 2010 article was written. For example, 2011 was a tragic year for tornado deaths and 2012 saw "record" heat and drought in the US followed by Superstorm Sandy. Sandy set a record for the lowest pressure ever

---

167. *Good Morning America*, "Al Gore: There is Still Time to Save the Planet" (June 23, 2006),
http://abcnews.go.com/GMA/GlobalWarming/story?id=2110628&page=2 (Accessed 6-8-10).
168. Charles J. Hanley, "Climate breakdown? U.N. chief less optimistic now" (August 13, 2010), *Associated Press*, *TheTimes-Tribune.com*, http://thetimes-tribune.com/news/health-science/climate-breakdown-u-n-chief-less-optimistic-now-1.944393 (Accessed 8-21-2010).

recorded in a hurricane to hit landfall north of Cape Hatteras, North Carolina.

Is it just me, or does that also strike you as an extremely lengthy description of a hurricane? It seems journalists worked overtime to find a way to describe Sandy as a *record* storm and an ominous "sign of troubling climate change already underway." Their work paid off, but Superstorm Sandy wasn't actually *super*. It didn't rank in the top ten of the lowest-pressure hurricanes in recorded history. The ten most severe hurricanes in order of intensity (low pressure) occurred in 1935, 1969, 2005, 1992, 1886, 1919, 1928, 1960, 1926, and 1961.[169] The northern route Sandy took before making landfall may have been an unusual path, but is that really enough to qualify as a superstorm?

The deadliest hurricane in US history occurred over 100-years ago in 1900, when a massive hurricane hit Galveston, Texas. Depending on which news report you reference, this storm killed 6,000 to 8,000 people. The hurricane was so powerful, it moved northeast out of Texas, crossed half the US Continent and entered the North Atlantic while maintaining its killer status. The second deadliest hurricane in US history was the Lake Okeechobee hurricane that hit Florida in 1928 killing 2,000. In 1893, the US suffered the fourth and the fifth deadliest hurricanes in its history.[170] In 1893, Earth had just moved out of the Little Ice Age and the US population was sparse compared to today, yet two of the five deadliest hurricanes in US history struck that year.

In November 2013, Supertyphoon Haiyan decimated the Philippines. Headlines described Haiyan as the most powerful storm in recorded history and global warming was generally cited as the cause of this "record" storm. However, the all-time record storm

---

169. Sara Gates, "The Most Intense Hurricanes In US History: Sandy Climbs Into The Ranks," *Huffington Post* (October 29, 2012), http://www.huffingtonpost.com/2012/10/29/most-intense-hurricanes-us-history_n_2041445.html#slide=1699757 (Accessed 11-14-12).

170. The Weather Underground, "Deadliest U.S. Hurricanes," (From the National Hurricane Center Publication, "The Deadliest Atlantic Tropical Cyclones, 1492 – Present"), http://www.wunderground.com/hurricane/usdeadly.asp (Accessed 11-19-12).

belongs to typhoon Tip (1979), which was the largest and the lowest pressure storm ever recorded. Tip spanned 1,380 miles—roughly the distance from New York to Dallas—covering an area of over 6,800 square miles.[171] By comparison, Haiyan spanned 300 miles—roughly the distance from Boston to Philadelphia—covering an area of less than 1,500 square miles.[172]

More importantly, between 1950 and 2013, more than a dozen typhoons in the same region registered lower pressure readings than Haiyan, including, Marge (1951), Nina (1954), Ida (1958), Nancy (1961), Emma (1962), Kit (1966), Amy and Irma (1971), and Nora (1973). The date of each of these nine typhoons is important because these *superstorms* all occurred between 1950 and the early 1970s—when global temperatures were falling. If global warming drives extreme weather, why were typhoons more severe during a period of global cooling? Also, why did two of the five deadliest US hurricanes occur near the end of the Little Ice Age and why did the deadliest hurricane in US history occur over 100-years ago?

The US drought of 2012 was called another record-setter by the mainstream media. As the drought grew in 2012, it was first called the worst drought since the 1980s, then the worst since the 1950s, and it was ultimately compared to the heat waves and drought that created the US Dust Bowl in the 1930s. This last comparison is absurd.

The Dust Bowl began in the summer of 1931 and lasted until the fall of 1939. This decade-long drought destroyed crops year after year after year and stripped away millions of tons of fertile topsoil. It was an environmental disaster of epic proportions, yet today's media insists on promoting the "record-setting" drought of 2012 as "a sign of troubling climate change already underway." Was

---

171. Megan Evans, "Earth's Strongest, Most Massive Storm Ever," *Accuweather.com* (October 16, 2012), http://www.accuweather.com/en/weather-news/typhoon-tip-earths-strongest-storm/87362 (Accessed 11-18-13).

172. Doyle Rice and Alia E. Dastagir, "Why everyone is talking about super typhoon," *USA Today* (November 8, 2013), http://www.usatoday.com/story/weather/2013/11/07/super-typhoon-philippines/3468165/ (Accessed 11-18-13).

the 2012 drought unusual, or would a longer-term view show that it was a routine event that occurs in the US every twenty to thirty years?

Extended droughts are also occurring in California and Texas. By mid-2014, the Texas drought was being described as the worst drought in 500-years. While we could certainly call that *unusual*, it's not unprecedented nor is it a sign of abnormal climate change because drought conditions were apparently worse 500 years ago.

As previously mentioned, 2011 was a tragic weather year with 529 tornado-related deaths in the US alone. It's easy to see why we fear the consequences of global warming, but the deadliest tornado in US history occurred on March 18, 1925. That's when the Tri-State Tornado killed 747 people across Missouri, southern Illinois, and Indiana. Another tornado outbreak on March 21, 1932 claimed 332 lives. Earlier, in 1917, tornadoes were responsible for 551 deaths. As far back as May 17, 1840 when scientists contend we were still in the grip of the Little Ice Age, a tornado struck Natchez, Mississippi, killing 317.

When we adjust the data to account for inflation and population growth, tornado damage has actually decreased since 1950, but we'll never see this fact in the mainstream media.[173] Are today's storms really record setters or does the media simply spin it that way? When we stop to consider the fundamental facts, we might realize that news anchors, weather forecasters, and journalists throw the word "record" around like a bunch of lonely frat-boys throwing beads at Mardi Gras.

Winter has also seen its share of extreme weather. On March 11, 1888, a blizzard called the Great White Hurricane dumped up to fifty inches of snow in the northeastern US. Winds created snowdrifts as high as fifty feet and as many as 400 people died. Two months earlier, the great School House Blizzard killed 230 in the Great Plains states. On January 27, 1922, the Knickerbocker Storm, covering some 22,000 square miles, dropped nearly three feet of snow. It collapsed the Knickerbocker Theater in Washington, DC, killing 98 and injuring 133. The world record for a single

---

173. Professor Roger Pielke Jr., "Hurricanes and Human Choice," *Wall Street Journal* (November 1, 2012) A17.

snowstorm is 189 inches of snow that fell at Mt. Shasta, California in 1959.

You might ask why anyone would mention record cold temperatures and snowfalls in a discussion of global warming. That's an excellent question. It's mentioned because we're now being told that human-caused global warming drove record-breaking cold temperatures in early 2014! Of course, it's no longer called *global warming*. It's called *climate change*, which is presumably why we can now blame greenhouse gases that warm the planet for causing record low temperatures.

The lows mentioned here occurred in early 2014 when the Eastern half of the US experienced repeated waves of frigid temperatures from the Canadian border all the way south to the Gulf Coast. We were told that global warming (*climate change*) was to blame.[174] The White House actually issued a blog and a video explaining why human-caused global warming caused the Arctic Vortex that created these record low temperatures.[175] *Time* magazine also ran an article blaming *climate change* for causing the record lows. Forty years earlier, however, *Time* ran a similar article blaming global *cooling* and the coming ice age for an Arctic Vortex that set record low temperatures in the Deep South during 1974.[176] Can we allow such double standards to dictate our beliefs?

Make no mistake: There's a reason the IPCC, environmentalists, and the media stopped using the term *global warming* and started using the term *climate change*. Despite their "undeniable proof," they can't guarantee that warming is the only

174. Andrew Freedman, "Polar Vortex in U.S. May be Example of Global Warming." *ClimateCentral.org*, January 6, 2014, http://www.climatecentral.org/news/polar-vortex-in-u.s.-may-be-valid-example-of-global-warming-16927 (Accessed 1-10-14).
175. Tom Bemis, "White House says 'polar vortex' likely caused by global warming," *Market Watch*, the *Wall Street Journal*, January 8, 2014, http://blogs.marketwatch.com/themargin/2014/01/08/white-house-says-polar-vortex-likely-caused-by-global-warming/ (Accessed 1-10-13).
176. Steven Goddard, "Time Magazine Goes Both Ways On The Polar Vortex," *Real Science*, January 7, 2014, http://stevengoddard.wordpress.com/2014/01/07/time-magazine-goes-both-ways-on-the-polar-vortex/ (Accessed 1-10-14).

change we'll experience over the coming decades and centuries. With this simple change in terminology, they can blame human activities for any and all changes in the weather—even if Earth moves into its next ice age. It's a subtle, yet brilliant change in terminology.

The phrase "climate change" is actually a redundant and completely meaningless choice of words. In all seriousness, blaming changing weather on "climate change" is like saying "the weather caused the weather to change." Earth's climate is always changing. There's another problem with "climate change." When it's mentioned, we generally interpret it to mean *human-caused* climate change. We've been trained to make the immediate connection, whether or not it's true.

Weather Proof?

Is today's weather abnormal, or have we allowed a sensationalizing media to convince us that it is? According to a 2013 article by Dr. Bjorn Lomborg, a founding member of one of the world's most prestigious environmental think tanks, the Copenhagen Consensus Center, **wildfires have decreased some 15% globally since 1950, global droughts have seen little change over the past sixty years, "hurricane activity is at a low not encountered since the 1970s [when Earth was cooling]," and "the US is currently experiencing the longest absence of severe landfall hurricanes in over a century**"[177] [emphasis added]. That's clearly not how the mainstream media or the environmental lobby describes today's weather. It's also the opposite of Al Gore's quote at the top of this chapter—a quote that allegedly serves as undeniable proof of human-caused global warming.

We believe today's storms are more severe than past storms because that's what we've been told to believe. We also develop short-term memories when it comes to weather. We forget the Dust Bowl and think the drought of 2012 was an unusual, human-caused event. Tornado deaths are always tragic, just as they were in 1840. We think Hurricane Sandy was a superstorm, but it was far less

---

177. Bjorn Lomborg, "Climate-Change Misdirection," *Wall Street Journal* (January 24, 2013), A15.

super than numerous storms that occurred during a period of global cooling and far less deadly than hurricanes that occurred near the end of the Little Ice Age. More importantly, we forget that "recorded history" for weather doesn't include records of the frequency or severity of storms during the Medieval Climate Optimum, the Little Ice Age, or any other past climate period.

Simply put, media stories sensationalizing today's weather could have been written for practically any year in Earth's history. The weather events happening today have all happened before. Today's extreme weather is not unique, abnormal, or any more dangerous than the extreme weather events of the past—despite sensational headlines to the contrary.

Chapter 12
Simple Things

"If the facts don't fit the theory, change the facts."
—Albert Einstein

One might believe that the simplest of all tasks involved in the study of global warming is comparing today's average global temperature to averages over the past 150 years. Unfortunately, this seemingly simple task is a bigger challenge than we might think. The IPCC warns that historic temperature records are plagued by changes in technology, changes in thermometer locations, and changes in thermometer exposure to the elements.[178] In other words, tracking historic changes in temperature is not an apples-to-apples comparison.

Our record of historic temperatures began in the late 1800s with as many as 1,000 thermometers widely scattered across Western Europe. By 1900, thermometers were readily available in many locations around the globe with the exception of the Poles where data wasn't available until the 1940s. The number of "official" recording stations grew to over 6,000 in the 1970s and 1980s, but has since declined to a total of only 2,400 sites.[179]

The lack of worldwide data prior to the 1940s and the constantly changing number and location of official temperature readings introduce uncertainties and the potential for errors and bias in the historic record.

---

178. J.T. Houghton, L.G. Meira Filho, B.A. Callander, N. Harris, A. Kattenberg and K. Maskell. "Climate Change 1995 The Science of Climate Change, Contributions of Working Group I to the Second Assessment Report of the Intergovernmental Panel on Climate Change," 181.
179. NASA, "GISS Surface Temperature Analysis, Station Data," http://data.giss.nasa.gov/gistemp/station_data/ (Accessed 12-19-12).

As an example, Skeptics warn that the majority of the 3,600 official stations that stopped tracking temperature after the 1980s were located in colder regions of the world. They also say many of the remaining official thermometers are increasingly impacted by the Heat Island Effect, which occurs when pavement, buildings, and other manmade structures absorb heat from the sun and radiate that heat to the surrounding atmosphere long after the sun has dropped over the horizon. Both of these issues would artificially increase "official" temperature readings taken since the 1980s.[180]

Believers say this is a non-issue because adjustments are made to account for the Heat Island Effect and corrections are applied to account for the loss of temperature data from colder regions. While there's no doubt that adjustments are being made, the fact remains; we can't be certain that the official temperature record is correct and unbiased. If we can't trust results from the seemingly simple task of tracking average temperatures, what data regarding global warming can we trust?

Global Cooling?

Skeptics have been criticized for claims that Earth has cooled since 1998. It's an effective way to discredit all Skeptics because it's presented as though the Skeptics won't even accept the obvious fact that Earth has warmed since the late 1800s. That's not what the Skeptics are saying.

There are actually three stories behind this alleged global cooling. The first story was driven by an unusually powerful El Nino, which peaked average surface temperatures in 1998. The years immediately following 1998 were cooler, but the warming has returned—according to official surface temperature readings.

The second story behind global cooling deals with IPCC climate models that predict the lower atmosphere should warm faster than Earth's surface. Satellite and balloon readings indicate the opposite

---

180. Joseph D'Aleo, Anthony Watts, "Surface Temperature Records: Policy-Driven Deception?" (August 27, 2010), 16. Space and Public Policy Institute, http://scienceandpublicpolicy.org/images/stories/papers/originals/surface_t emp.pdf (Accessed 12-12-12).

has occurred. This tells us one of four things: A) The IPCC computer models are wrong; B) The ground-based temperature readings are wrong; C) The satellite and balloon readings are wrong; or D) All of the above.

In this case, Believers say the satellite and balloon readings are wrong. It's true! When temperature readings from the lower atmosphere didn't support the IPCC models, NOAA made corrections to the temperature data. As satellites age, their orbits can slow, so corrections were made to account for possible changes in the time of day when temperature readings were taken. This resulted in an upward adjustment to the latest satellite readings. The balloon thermometers used today are better protected from direct sunlight than older thermometers, so corrections were also made to lower earlier temperature readings to account for the impact of direct sunlight on those thermometers. Together, these two adjustments brought temperature readings into closer agreement with the IPCC computer models.[181]

While there's nothing wrong with NOAA making needed corrections to their data, Skeptics wonder why the need for these corrections didn't surface until someone questioned the disagreement with IPCC model projections.

The third story regarding global cooling was highlighted in a September 2013 *Los Angeles Times* article titled "Global warming 'hiatus' puts climate change scientists on the spot."[182] The article discussed the IPCC's fifth assessment report, which indicates that warming since 1998 has been far below even the most optimistic IPCC computer projections. The article anticipates an upcoming "heated debate" among IPCC scientists regarding this lack of warming. Included in the article is a quote from a past member of the IPCC, Judith Curry, who heads the School of Earth and

---

181. Phil McKenna, "Climate myths: The lower atmosphere is cooling, not warming" (May 16, 2007), http://www.newscientist.com/article/dn11660-climate-myths-the-lower-atmosphere-is-cooling-not-warming.html (Accessed 12-8-12).
182. Monte Morin, "Global warming 'hiatus' puts climate change scientists on the spot," Los Angeles Times September 22, 2013, http://articles.latimes.com/2013/sep/22/science/la-sci-climate-change-uncertainty-20130923 (Accessed 12-16-13).

Atmospheric Sciences at the Georgia Institute of Technology. After accusing the IPCC of intellectual arrogance and bias, Curry concluded by saying ". . . what we are seeing is natural climate variability dominating over human impact[s]." Francis Zwiers, a vice chair of the IPCC, indicated the IPCC computer models have "significantly" overestimated global warming for the last 20-years. *Nature* covered Professor Zwiers comments in an article titled "Overestimated global warming over the past twenty years."[183]

In an earlier chapter, I mentioned an annual New Year's Day round of golf as proof of global warming. Now that I think about it, it's been too cold the last few years to enjoy that annual outing. Is it possible global warming has slowed or stopped despite increased greenhouse gas emissions?

Expanding Glaciers

Melting glaciers are commonly used to prove that global warming is real. What would it mean if glaciers were now beginning to grow? It's a fact. Many glaciers around the globe have stopped retreating and are now beginning to expand, including glaciers in Alaska, California, and the Himalayas.[184, 185, 186, 187] Does that mean

183. John C. Fyfe,   Nathan P. Gillett & Francis W. Zwiers, "Overestimated global warming over the past 20 years," published online August 28, 2013,
http://www.nature.com/nclimate/journal/v3/n9/full/nclimate1972.html (Accessed 12-17-13).
184. *Discovery News, Earth*; "Himalayan Glaciers Seem to be Growing," (February 11, 2013), http://news.discovery.com/earth/himalayas-glaciers-shrink.htm (Accessed 5-5-13).
185. *Iceagenow.com*, "Glaciers are growing around the world, including the United   States,"   http://www.iceagenow.com/Growing_Glaciers.htm (Accessed 7-10-10).
186. *Associated Press* (July 9, 2008), "Mysterious California Glaciers Keep Growing Despite Warming"
http://www.foxnews.com/story/0,2933,378144,00.html   (Accessed   11-14-11).

global warming has ended?

Believers have a ready-made answer to explain why expanding glaciers can also indicate global warming. Their explanation indicates that increased temperatures increase ocean evaporation, which leads to increased rainfall, and at higher elevations, the increased rainfall turns into greater snowfall. The increased snowfall accumulates on existing glaciers, causing them to expand.

That's a reasonable argument, but it smacks of a double standard. Whether glaciers are shrinking or expanding, it's proof that Earth is warming. If satellite data doesn't agree with the IPCC, correct the satellite data to prove that Earth is warming. Every time the weather changes, be it extreme heat and drought or excessive rain and flooding, it's proof of global warming. If we experience record low temperatures like those of early 2014, it's proof that our planet is getting hotter.

It seems Albert Einstein was right: "If the facts don't fit the theory, change the facts." How many double standards are we willing to accept before we begin to question the strength of such arguments?

The Poles Are In

Mainstream media reports continue to track the rapid melting of Arctic sea ice. In late 2013, the media referred to the volume of Arctic ice as the sixth lowest ever recorded during October. What they seldom mentioned, however, was that this level of Arctic ice was some 50% greater than in October 2012.[188] Numerous yachts

---

187. D.C Trabant, R.S. March, and D.S. Thomas, D.S.;"Hubbard Glacier, Alaska: Growing and Advancing in Spite of Global Climate Change and the 1986 and 2002 Russell Lake Outburst Floods," US Geologic Survey Fact Sheet (2003),
http://ak.water.usgs.gov/glaciology/hubbard/reports/200301_fs001-03/index.htm (Accessed 11-14-11).
188. Steve Conner, "The good news: there has been a dramatic increase in Arctic sea ice. The bad: it's still half the level it was in the 1980s," December 16, 2013, *The Independent*,
http://www.independent.co.uk/environment/climate-change/the-good-news-there-has-been-a-dramatic-increase-in-arctic-sea-ice-the-bad-its-still-half-the-level-is-was-in-the-1980s-9008388.html (Accessed 12-16-13).

that were traversing the Arctic waters found themselves trapped by the ice and forced to wait until the summer of 2014 to complete their journeys. In August 2010, satellites recorded the world's lowest naturally occurring temperature of all time, minus 135.8 degrees Fahrenheit. This record low was nearly matched by the minus 135.3 degrees Fahrenheit reached on July 31, 2013.[189] Also in 2013, the extent of Antarctic sea ice was the highest ever recorded.[190] The US set more record lows than record highs in 2013[191] and 2014 began with widespread record lows across half the US.

Are we certain that CO_2 is the primary driver of Earth's climate and that global warming is the only climate change we'll see in the future?

A New Ice Age?

Despite recent temperatures described as "the highest ever recorded," many scientists say another ice age may be lurking in the near future. A November 1, 2009 article in *Pravda* by Gregory F. Fegel titled "Earth on the Brink of an Ice Age" discusses a growing level of scientific data from climatology indicating that Earth is nearing its next ice age.[192]

If you have any doubts, search the web for "CO_2 is preventing the next ice age." You'll find strong agreement that solar activity is set to enter a period of low intensity similar to that experienced during past cold periods and ice ages. You'll find indications that recurring changes in Earth's orbit and repeating cycles of cold water

---

189. *CBS News/AP (Associated Press)*, "Antarctic temperature hit record low," December 10, 2013, http://www.cbsnews.com/news/antarctic-temperature-hit-record-low/ (Accessed 12-16-13).

190. Guy Williams, "Why is Antarctic sea ice growing?," October 29, 2013, Physics.org (PHYS.ORG), http://phys.org/news/2013-10-antarctic-sea-ice.html, (Accessed 12-12-13).

191. Doyle Rice, "Cold facts: More record lows than highs in USA in 2013," *USA Today*, January 2, 2014,
http://www.usatoday.com/story/weather/2013/12/31/record-cold-temperatures/4264237/ (Accessed 1-13-14).

192. Gregory F. Fegel, "Earth on the Brink of an Ice Age" (November 1, 2009); *Pravda.ru*; 1, http://english.pravda.ru/science/earth/106922-earth_ice_age-0 (Accessed 7-10-10).

circulating from ocean depths to the surface will soon reverse their warming influence and begin cooling the planet. You'll also find studies warning that we don't have enough fossil fuel to stop the next ice age once conditions change to induce cooling rather than warming.

The studies you'll find suggest that our use of fossil fuels will delay the onset of the next ice age. That would reinforce everything we've grown to believe about the role of CO_2 in warming our planet with one exception. If our CO_2 emissions are helping Earth avoid the next ice age, isn't that good news regarding those emissions?

Solar System Warming

Prior to the IPCC's fifth assessment report acknowledging that Earth's warming has slowed, NASA satellites were indicating that Mars; Jupiter; Pluto; and Neptune's moon, Triton, were experiencing warming similar to Earth. Clearly our emissions of CO_2 have nothing to do with temperatures on other planets. It seems reasonable to assume that whatever is impacting temperatures on nearby planets would also impact Earth. Perhaps it's the sun?

A 2007 article from Canada's *National Post* discusses this solar system warming. The article indicates that Pluto and Jupiter are warming and frozen nitrogen is melting on Neptune's moon, Triton.[193] *National Geographic* ran a story on October 28, 2010 titled "Mars Melt Hints at Solar, Not Human, Cause for Warming, Scientist Says."[194] The title speaks for itself.

Search the Web for "warming solar system" or "NASA warming

---

193. Lorne Gunter; "Warming On Jupiter, Mars, Pluto, Neptune's Moon & Earth Linked to Increased Solar Activity, Scientists Say," Taken from "Bright Sun, Warm Earth. Coincidence?" Senate Committee on Environment and Public Works, published in *Canada's National Post* (March 12, 2007),
http://epw.senate.gov/public/index.cfm?FuseAction=PressRoom.Blogs&ContentRecord_id=469DD8F9-802A-23AD-4459-CC5C23C24651 (Accessed 7-7-10).
194. Kate Ravilious, "Mars Melt Hints at Solar, Not Human, Cause for Warming, Scientist Says," *National Geogrpahic.com* (February 28, 2007) http://news.nationalgeographic.com/news/2007/02/070228-mars-warming.html (Accessed 5-7-11).

planets," and you'll find NASA reports indicating that our solar system is indeed warming.[195] Some articles suggest that Mars is warming so fast it's in danger of losing its polar ice caps.[196, 197] Sound familiar?

Simple Things

We perceive a long drought as a sign of impending doom, but only because we've allowed the media to spin it that way. We believe increasing temperatures demonstrate abnormal climate change driven by human greenhouse gas emissions, yet the IPCC has told us that Earth was recently warmer than today despite lower CO_2 levels. Superstorm Sandy wasn't all that super. We didn't know that large and abrupt climate changes are routine events. We didn't realize that ocean levels have changed dramatically in the past and today's increases are nothing new. No one ever told us that today's temperatures are below the average temperature of the last 10,000 years. No one told us that nearby planets are warming. The media largely ignored reports of cooling since 1998 and reports indicating that any warming since that time has been far below IPCC predictions.

We haven't been allowed to see simple facts that would dramatically change our view of today's climate. Instead of presenting these facts, the media has either ignored the simple things discussed here or presented them as dangerous and abnormal events—proof that this time, climate change is our fault.

Perhaps the most important fact that has been ignored is also the simplest: Recorded history for temperature began at the end of the Little Ice Age. Knowing that, shouldn't we expect today's temperatures to be the highest ever recorded?

---

195. James M. Taylor, "Mars Is Warming, NASA Scientists Report," *Environment & Climate News* (November 1, 2005), http://www.heartland.org/policybot/results/17977/Mars_Is_Warming_NASA _Scientists_Report.html (Accessed 7-16-11).
196. Jonathan Leake, "Climate Change Hits Mars," (April 29, 2007) *The Sunday Times*, http://www.sott.net/article/131146-Climate-Change-hits-Mars (Accessed 8-12-08).
197. Lome Gunter, et al.

Chapter 13
Pants on Fire

"People will believe a big lie sooner than a little one and if you repeat it frequently enough, people will sooner or later believe it."
—Walter Langer

Walter Langer was the psychoanalyst who developed the top secret personality profile of Adolf Hitler in World War II. He accurately predicted many of Hitler's actions during the war, including Hitler's suicide at the end of the war. Langer's work led to the creation of the FBI's Behavioral Analysis Unit (BAU), which is highlighted in the television drama, *Criminal Minds*.

According to Langer, the bigger the lie, the more likely we are to believe it, and if the lie is repeated often enough, we'll eventually accept it as the truth. Is it possible that Langer's analysis of human behavior explains our belief in human-caused climate change?

Global Warming's Holy Grail

If there was a rallying point for Believers, it was the 2006 release of the documentary film, *An Inconvenient Truth*—the Holy Grail of global warming. The film leaves no doubt: Earth is on the brink of destruction and humans are responsible. This was a powerful and influential film. It was powerful enough to win an Oscar for best documentary, and influential enough to earn Al Gore a share of the Nobel Peace Prize. Many who were previously skeptical of humanity's influence on climate became Believers after seeing the film. Was their change in beliefs justified?

The Film's Strongest Evidence

We need look no further than what many call the film's "strongest evidence" to realize this film is filled with intentional deception. The strongest evidence that humans cause climate change was shown in two graphs. The two graphs charted the long-term history of Earth's temperatures and CO_2 levels. As shown in the film, it's clear that every time CO_2 increased, temperature increased; and every time CO_2 decreased, temperature decreased. Case closed: Changes in CO_2 drive changes in temperature!

Houston, we have a problem. The historic data used to create these two graphs actually show that temperatures have always increased before CO_2 increased and temperatures have always declined before CO_2 declined! In fact, history shows that temperatures began rising even though CO_2 levels were decreasing and temperatures began declining even though CO_2 was increasing.[198, 199, 200] Something other than CO_2 has always caused temperature to change first.

That's clearly not the message delivered by the film. The fundamental facts tell us this documentary was intentionally deceptive. The deception was easy to create.

Because the timespan shown in the graphs was so large (650,000 years), viewers couldn't see what was happening over the course of a few centuries' time. Had Mr. Gore or the film's producers zoomed-in on a smaller time scale, viewers would have seen the actual inconvenient truth: CO_2 has always been a climate follower, not a climate driver. Figure 13-1 on the next page is a rough recreation of what's actually occurring.

---

198. Fischer, H., Wahlen, M., Smith, J., Mastroianni, D. and Deck, B. (1999), "Ice core records of atmospheric CO2 around the last three glacial terminations," *Science*, 12 March 1999: Vol. 283 no. 5408, 1712-1714.

199. Petit, J.R., Jouzel, J., Raynaud, D.,Barkov, N.I., Barnola, J.-M., Basile, I., Bender, M., Chappellaz, J., Davis, M., Delaygue, G., Delmotte, M., Kotlyakov, V.M., Legrand, M., Lipenkov, V.Y., Lorius, C., Pepin, L., Ritz, C., Saltzman, E. and Stievenard, M., (1999); "Climate and atmospheric history of the past 420,000 years from the Vostok ice core, Antarctica," *Nature* 399: 429-436.

200. Staufer, B., Blunier, T., Dallenbach, A., Indermuhle, A., Schwander, J., Stocker, T.F., Tschumi, J., Chappellaz, J., Raynaud, D., Hammer, C.U. and Clausen, H.B. 1998, "Atmospheric CO2 concentrations and millennial-scale climate change during the last glacial period," *Nature* 392: 59-62.

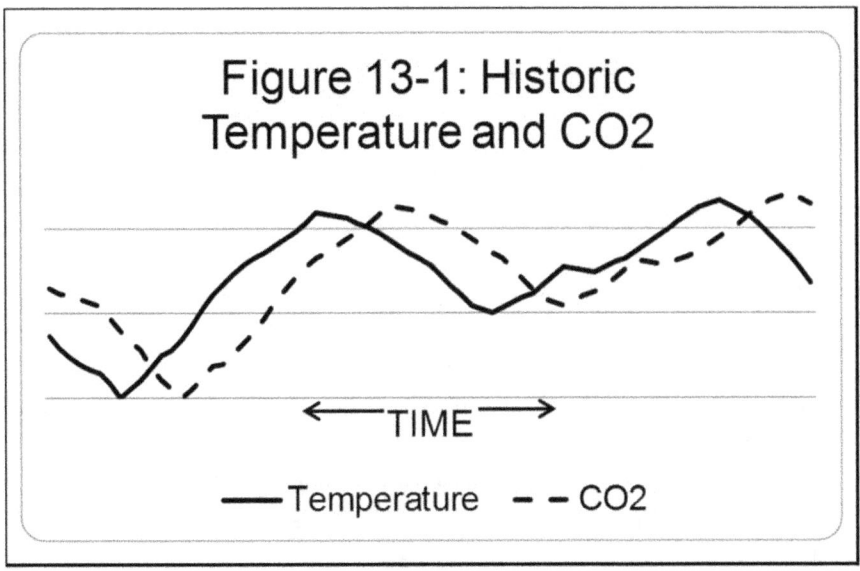

Figure 13-1: Historic Temperature and CO2

——Temperature   – – CO2

Following the solid line (temperature) from the left, you'll see that temperatures are initially declining, but soon reverse direction and begin rising. If you follow the dashed line (CO_2 levels) from the left, you see it does the same thing with one exception: It continues to decline after temperatures begin to increase. If we continue to follow both lines, we see that temperature once again reverses direction and begins to fall, even though CO_2 is still increasing. Over the entire timespan shown in the film, temperatures always changed direction first and CO_2 always followed. It never happened the other way around![201]

The reason CO_2 levels change based on changes in temperature is easily explained. Earth's oceans are our planet's largest carbon *sinks*. In other words, oceans capture and store more CO_2 than plants or land masses. The ocean's ability to capture and store CO_2 is controlled by numerous variables, but one of the primary variables is temperature. Cold ocean waters capture and store large volumes of CO_2. Conversely, as oceans warm, they begin releasing CO_2. This simple reality is well understood. Many

---

201. John Stossel, "Is there really a global warming consensus?" *ABC 20/20*, http://www.youtube.com/watch?v=BZcp_wcDXec (Accessed 11-11-08).

scientists contend that because our planet is warming as natural events drive Earth's recovery from the Little Ice Age, atmospheric CO_2 would be increasing even if humans weren't burning fossil fuels.

The film's editors had to realize that their own data showed that CO_2 follows temperature, but they went out of their way to ensure the audience didn't see this inconvenient truth. It was an unconscionable breach of public trust. The film's strongest evidence demonstrates that something other than CO_2 has always been the primary driver of climate change and CO_2 has always been a climate follower—yet the film intentionally left the opposite impression.

What's just as troubling as the film's intentional deception is the fact that the media continues to use these same graphs to reinforce the belief that CO_2 drives climate change.

Lake Chad

Satellite pictures of Lake Chad were another influential set of images shown in the film. The images showed the lake disappearing over the course of only a few years' time. The film blames human-caused global warming. The fundamental facts point to other reasons.

Lake Chad is a shallow lake with an average depth of less than five feet. Its deepest point is a mere thirty-four feet.[202] Any extended period of dry weather will dramatically shrink such a shallow lake over a short period of time. Conversely, any extended wet period will quickly refill the lake and extend its boundaries. In any one-year period, such a shallow lake could substantially grow or shrink.

Throughout history, the surface area of this lake has changed dramatically. At times it nearly disappeared, and at other times it covered 154,000 square miles.[203] These past changes were driven by routine, natural changes in the weather, not human emissions of greenhouse gases.

---

202. *Princeton.edu*, "Lake Chad,"
http://www.princeton.edu/~achaney/tmve/wiki100k/docs/Lake_Chad.html (Accessed 10-21-13).
203. Ibid.

In truth, human activities are causing Lake Chad to shrink, but those activities have nothing to do with CO_2 or global warming. They're related to a booming regional population that is increasingly using the lake for drinking water and irrigation. Lake Chad wasn't shown because it proves humans cause global warming. It was shown because—like the film's strongest evidence—it delivered the desired message.

Banned from Classrooms?

When public schools in the UK planned to show *An Inconvenient Truth* to students, at least one parent objected. Steward Dimmock, the father of two school-age children, filed suit to block use of the film in classrooms. He saw this film as political indoctrination, not sound science. UK law prohibits political indoctrination in the classroom.

The UK's High Court agreed that the film was political indoctrination, not a scientific work. The court stopped short of banning the film and instead ruled that *An Inconvenient Truth* could not be shown in classrooms unless teachers first read a disclaimer.[204] Teachers were required to say the film is a political work, not a scientific work, and they had to read a list of claims made in the film that were not supported by science.[205] Depending on which article you reference, the number of inaccuracies teachers must identify is either nine or eleven. Noel Sheppard and *NewsBusters.org* listed eleven inaccuracies that must be cited by teachers before showing the film. In that list, shown below, the term *government's expert* refers to various subject-matter experts defending the accuracy of *An Inconvenient Truth*:

---

204. Lewis Smith, "Al Gore's An Inconvenient Judgment," *Times Online*, October 11, 2007,
http://www.uio.no/studier/emner/matnat/ifi/MNSES9100/h11/undervisnings
materiale/literature/Al%20Gore%E2%80%99s%20inconvenient%20judgem
ent%20times%202007.pdf (Accessed 3-23-10).
205. Noel Sheppard, "Court Identifies Eleven Inaccuracies in Al Gore's 'An Inconvenient Truth'" (10-9-07), *NewsBusters.com*,
http://newsbusters.org/blogs/noel-sheppard/2007/10/09/court-identifies-
eleven-inaccuracies-al-gore-s-inconvenient-truth (Accessed 8-1-09).

1. The film claims that melting snows on Mount Kilimanjaro [are] evidence [of] global warming. The Government's expert was forced to concede that this is not correct.
2. The film suggests that evidence from ice cores proves that rising CO_2 causes temperature increases over [the most recent] 650,000 years. The Court found that the film was misleading: over that period the rises in CO_2 lagged behind the temperature rises [In other words, something other than CO_2 has always driven temperature changes].
3. The film uses emotive images of Hurricane Katrina and suggests that this has been caused by global warming. The Government's expert had to accept that it was 'not possible' to attribute one-off events to global warming.
4. The film shows the drying up of Lake Chad and claims that this was caused by global warming. The Government's expert had to accept that this was not the case.
5. The film claims that a study showed that polar bears had drowned due to disappearing arctic ice. It turned out that Mr. Gore had misread the study: in fact four polar bears drowned and this was because of a particularly violent storm.
6. The film threatens that global warming could stop the Gulf Stream throwing Europe into an ice age: the Claimant's evidence was that this was a scientific impossibility.
7. The film blames global warming for species losses, including coral reef bleaching. The Government could not find any evidence to support this claim.
8. The film suggests that the Greenland ice covering could melt causing sea levels to rise dangerously. The evidence is that Greenland will not melt for millennia.
9. The film suggests that the Antarctic ice covering is melting; the evidence [presented in court, but not the film] was that it is in fact increasing.
10. The film suggests that sea levels could rise by 7m [meters] causing the displacement of millions of people. In fact the evidence is that sea levels are expected to

rise by about 40cm over the next hundred years and that there is no such threat of massive migration.

11. The film claims that rising sea level has caused the evacuation of certain Pacific islands to New Zealand. The Government are [*sic*] unable to substantiate this and the Court observed that this appears to be a false claim.

Search the web for "UK court bans Gore's film" and you'll see numerous entries on this topic. Your computer search might also reveal websites discrediting Mr. Dimmock, who is a common citizen and father who provides for his family by driving a lorry. It's a common tactic: Anyone challenging the popular theory of global warming becomes a target for ridicule. One such claim indicated that a network of energy providers financed Mr. Dimmock's lawsuit. Apparently, we're supposed to ignore the Court's ruling because Mr. Dimmock was "on the take." I can't tell you if this claim is true or not, but it leads to an important question. How would Mr. Dimmock's finances influence the Court's decision or change the significance of their findings? Does it really matter who funded the court case?

For many of us, it does matters. If Big Energy was involved, it must be a lie, right? That's what we've been trained to believe, but there's no evidence the Court was "on the take." Facts were presented, evidence was heard, and Gore's film lost—regardless of Mr. Dimmock's financial situation or who provided the funds to pursue the case.

This film is not a work of science. It's little more than an environmental infomercial; perhaps the longest infomercial in history. It's essentially a marketing gimmick backed by Hollywood and the environmental lobby. The marketing goal had terrific intentions. It was intended to sway our beliefs so we would change our energy-related behaviors. While we can all agree with the need to change energy behaviors, does that justify the film's obvious deceptions?

This film wasn't designed to inform. It was designed to deceive—and it was a rousing success. Instead of being upset by Mr. Dimmock's finances, shouldn't we be far more upset by an influential film that was misleading and, apparently, intentionally misleading?

The real irony is that the UK High Court decision came one day before Al Gore was announced as a co-winner of the Nobel Peace Prize for his role in the film.

Courtroom Testimony

One scientist testifying in opposition to the film's use in classrooms was Professor Robert Merlin Carter of James Cook University, Queensland, Australia. Professor Carter is a paleontologist, marine geologist, and environmental scientist with over forty years of experience. He holds tenured positions at more than one leading university, he was the Australian Council Chairman of the Earth Science Departments, and has served as a member of the Australian Senate Select Committee on Climate Policy. Highlights of Professor Carter's testimony are listed below:[206]

- Nowhere does Mr. Gore tell his audience or readers that all of the phenomenon described in *AIT* [*An Inconvenient Truth*] fall within the natural range of previous environmental change on our planet.
- [The film neglects to tell the audience] That hot years clustered around the end of the 20th century is [*sic*] irrelevant because of the limited context provided by considering only 150 years worth of data. 150 years is a short and climatically almost meaningless period.
- It [the film] comprises a catalogue of circumstantial and anecdotal [*sic*] evidence that is remorselessly cherry-picked to exemplify worst-case warming scenarios; at the same time, Mr. Gore neglects altogether the undoubtedly real risks of future climatic cooling.
- It [the film] selectively omits discussion of aspects of the climate system that do not reinforce the villainization [*sic*] of human-produced greenhouse gases; for example both

---

206. "Second Witness Statement of Professor Robert Merlin Carter" (August 22, 2007), High Court of Justice, Queens Bench Division, Administrative Court, Claim CO/3615/2007, http://newparty.co.uk/sites/all/themes/newparty/files/carterstatement.pdf (Accessed 1-10-10).

the major source of climate energy and variation (solar) and the major atmospheric greenhouse gas (water vapour) are ignored.

- It is important to point out that the IPCC is itself an arm of the UN and therefore part of a political organization. Its panel members are selected by politicians and it has an agenda of its own. Several polls of professional opinion indicate that widespread scientific scepticism [sic] exists with the IPCC orthodoxy.

- The statement [in the film], in context implies that CO2 is a pollutant. This is incorrect. Rather, in scientific terms, CO2 is a colourless, odourless natural gas that we breathe out every time we exhale, that is an essential constituent of plant growth, and that has been present in earth's atmosphere through time in trace amounts ranging from a few hundred to many thousand ppm.

- This [computer model] uncertainty is acknowledged by the IPCC (2001, 2007), which deliberately uses the term "scenario" rather than "prediction" for its GCM-forecasted [computer-forecasted] futures. Unfortunately, the press and public (including Mr. Gore) are not so careful, and generally treat IPCC scenarios as firmly predicted futures.

We could continue listing Professor Carter's arguments, but there's little more to be gained. The bottom line is evident: *An Inconvenient Truth* presents a biased, prejudicial, and completely one-sided view of global warming—just like the mainstream media.

A Biased Media

One article tells us all we need to know about the media's bias related to global warming and environmental issues. The article in question is titled: "Man loses bid to ban An Inconvenient Truth from

U.K. schools."[207] The title speaks for itself. The article leads readers to believe that Mr. Gore's documentary has withstood all challenges and has been upheld as the gospel.

Yes, this media headline is absolutely true, but it's clearly misleading. It's difficult to see how anyone could describe this article as anything other than intentionally deceptive. It seems the author and editors of this article were predisposed to cherry-pick their facts to support the popular theory of global warming.

Walter Langer was right: "People will believe a big lie sooner than a little one and if you repeat it frequently enough, people will sooner or later believe it."

---

207. "Man loses bid to ban An Inconvenient Truth from U.K. schools," *CBC news online* (October 11, 2007), http://www.cbc.ca/news/arts/film/story/2007/10/11/inconvenient-truth.html (Accessed 4-24-12).

Chapter 14
CarbonGate

"You can observe a lot by watching."
—Yogi Berra

Prior to the rise of the global warming issue, we had few opportunities to *observe* any news related to CO_2. There's a reason: CO_2 is not a pollutant. Carbon dioxide may be a heat-trapping greenhouse gas, but it occurs naturally and it's as essential for life on this planet as water and oxygen.[208, 209] That's clearly not how CO_2 is being portrayed today.

As an example, there's a CO_2-related public service advertisement that clones a popular phone company commercial. In the CO_2 ad, an adult is sitting at a low table with several preschoolers. The adult asks an obvious question similar to the following: "What's better, CO_2 and debilitating asthma or no CO_2 and no asthma?" It's a great commercial except for the fact that CO_2 doesn't cause asthma. In order to connect CO_2 to asthma, we'd have to connect a lot of dots. We'd have to argue that CO_2 increases temperatures; high temperatures are required for the formation of smog; smog can trigger symptoms of asthma in those with the disease; therefore, CO_2 helps trigger asthma. Okay, so there is a way to connect the dots, but it's a precarious connection

---

208. "Carbon Dioxide (CO2) is Not a Pollutant," (November 20, 2008), *Popular Technology.net*;
http://www.populartechnology.net/2008/11/carbon-dioxide-co2-is-not-pollution.html (Accessed 2-14-10).
209. William Happer, "CO2 is Not a Pollutant: Debunking a Global Warming Myth," *Witherspoon Institute*, Public Discourse (December 1, 2009), http://www.thepublicdiscourse.com/2009/12/1037 (Accessed 2-14-10).

at best. The commercial typifies the way CO_2 is being portrayed in the media today. Is it a fair portrayal?

If we take a similarly precarious path, we could also say that CO_2 will reduce asthma attacks. How is that possible? If we remember the claim that expanding glaciers can be caused by global warming, we'll remember that CO_2 increases evaporation and rainfall. Therefore, we could say that CO_2 reduces the number of asthma attacks because it leads to more rainfall, which removes asthma triggers from the air. Accepting one argument and ignoring the other would require us to apply yet another double standard.

If we insist on calling CO_2 a pollutant, we have to realize that it's like no pollutant we've ever known. It clearly benefits plants, which sit at the bottom of the food chain. That's a good thing, not a bad thing. CO_2 is not responsible for dirty air, lung disease, heart disease, blood disorders, smog, arthritis, cancer, warts, or any other human health concern. We can address asthma, energy independence, air pollution, water pollution, improved health, increased standards of living, and yes, even landfills without ever addressing CO_2. We can fight deforestation, save the ozone layer, and save the whales—and never address CO_2. If we call CO_2 a pollutant, perhaps we need a new definition for the term *pollutant*.

For those insisting that CO_2 is a pollutant, the EPA has come to the rescue. As previously discussed, the EPA has determined that CO_2 endangers public health due to global warming. Therefore, we can now officially call CO_2 a pollutant—even though it's like no pollutant we've ever known.

There's also more ammunition for those wishing to call CO_2 a pollutant. Two new environmental evils associated with CO_2 are ocean acidification and coral bleaching. We'll discuss these two issues at the end of this chapter, but let's first focus on the primary reason we call CO_2 a pollutant: its alleged role in climate change.

Our Fragile Atmosphere

We're told that if Earth were the size of a normal classroom globe, our atmosphere would appear as a thin layer of varnish covering that globe. With this analogy, it's easy to understand why daily emissions of 90 million tons of anything could be considered a pollutant. This perception, however, changes when we consider the full-scale model of our atmosphere.

Earth's atmosphere weighs in at a hefty 5.5 quadrillion tons. On an annual basis, our daily 90 million tons of CO_2 adds about one molecule to the atmosphere for every 500,000 existing molecules, or roughly 0.0002%.[210] Additionally, land ecosystems and oceans adsorb some of those emissions. Our contributions are far smaller than we're led to believe.

Natural Sources of CO_2

While humans are emitting 90 million tons of CO_2, Mother Nature is emitting over 1.8 billion tons: more than 95% of all CO_2 emissions.[211] Therefore, Skeptics argue that if CO_2 is causing any problems, humans are responsible for less than 5% of those problems. Believers, of course, disagree. They say that human emissions upset Earth's natural balance, meaning humans are responsible for all of the issues caused by CO_2.

Both arguments seem reasonable, but as always, there's more to the story. Carbon dioxide isn't the only greenhouse gas emitted into the atmosphere—and not all greenhouse gases are created equal.

Global Warming Potential (GWP)

Some greenhouse gases cause more warming than others. For example, one ton of methane causes twenty to twenty-four times more warming than one ton of CO_2. Therefore, the Global Warming Potential (GWP) of methane is twenty to twenty-four times greater than that of CO_2. Of the nearly forty different greenhouses gases studied by the IPCC, CO_2 has the lowest GWP on a pound-for-

---

210. Dr. Roy W. Spencer, "Climate Confusion: How Global Warming Hysteria Leads to Bad Science, Pandering Politicians, and Misguided Policies that Hurt the Poor," 63.
211. *EcoInfo.net*, "Carbon Dioxide: A Greenhouse Gas," http://www.eco-info.net/carbon-dioxide-a-greenhouse-gas.html (Accessed 12-28-12).

pound basis.[212] It's the volume of CO_2 in the atmosphere that allegedly creates the concern.

Water Vapor

Depending on the study referenced, water vapor is responsible for 60% to 95% of the total greenhouse effect. The most widely accepted value for water vapor's portion of the greenhouse effect is 95%.[213] Based on this figure, if we consider all the greenhouse gases in the atmosphere and we account for their GWP, human contributions of CO_2 are responsible for 0.5% of the total greenhouse effect. This may or may not change your view of humanity's role in global warming, but it's a seemingly important fact we haven't heard from the mainstream media.

Did you know the UN's plan to save the planet from global warming (the Kyoto Protocol) targets less than half of all human greenhouse gas emissions? The UN ignores emissions from developing countries even though those countries emit over half of all human-related greenhouse gases. Based on the UN plan, all of our fears and all of our actions to save the planet are focused on reducing—not eliminating, but simply reducing—less than 0.25% of the total warming impact tied to greenhouse gases! This is another seemingly important fact missed by the mainstream media.

Many scientists contend that our efforts to reduce CO_2 emissions will have no meaningful impact on our climate. We are, after all, targeting a very small percentage of the greenhouse effect. Still, Believers insist that our small contribution upsets Earth's natural balance of CO_2. That's easy enough to accept, but what is Earth's natural balance of CO_2?

---

212. Solomon, S., D. Qin, M. Manning, Z. Chen, M. Marquis, K.B. Averyt, M. Tignor and H.L. Miller (eds.), "Contribution of Working Group I to the Fourth Assessment Report of the Intergovernmental Panel on Climate Change, 2007" Cambridge University Press, Cambridge, United Kingdom and New York, NY, USA. "Climate Change, The Physical Science Basis, Working Group I," Table TS.2. 33 (Accessed 6-11-12).
213. *Thinkglobalgreen.org*, "Water Vapor" (March 21, 2013), http://www.thinkglobalgreen.org/WATERVAPOR.html (Accessed 10-22-13).

Historic CO_2 Levels

Many different approaches have been used in an effort to determine historic levels of atmospheric CO_2 and to recreate Earth's historic temperatures. Each of these different approaches yields slightly different results, but they all agree on one conclusion: **Today's level of atmospheric CO_2 is very low by historic standards**. I've found no scientific disagreement on this issue, only various efforts to explain why we should ignore this historic fact.

There are numerous studies documenting high past levels of CO_2. An article in *Science*, indicates that between 25 and 45 million years ago, CO_2 concentrations ranged from 1,000 ppm (parts per million) to 1,500 ppm. [214] That's some two and a half to four times higher than today, but still very low by historic standards.

A more detailed view of carbon's history comes from NASA's Goddard Space Flight Center and data from the GEOCARB III study. This study recreated CO_2 levels over the past 600 million years. Figure 14-1 on the next page shows the results. It shows that Earth's atmosphere has contained some twenty-five times more CO_2 than it contains today. It also shows that we'd have to go back roughly 300 million years to find CO_2 levels as low as those Earth has experienced over the past several hundred thousand years. [215]

214. Mark Pagani, James C. Zachos, Katherine H. Freeman, Brett Tipple, Stephen Bohaty, "Marked Decline in Atmospheric Carbon Dioxide Concentrations During the Paleogene" (January 21, 2005, Published Online June 16, 2005), *Science* 22 July 2005: Vol. 309 no. 5734 600-603 DOI: 10.1126/science.1110063,
http://www.sciencemag.org/content/309/5734/600 (Accessed 9-2-11).
215. R.A. Berner and Z. Kothavala, "GEOCARB III: A Revised Model of Atmospheric CO2 over Phanerozoic Time," http://s155.n46.n171.n68.static.myhostcenter.com/WVFossils/Reference_Docs/Geocarb_III-Berner.pdf, (Accessed 8-31-11).

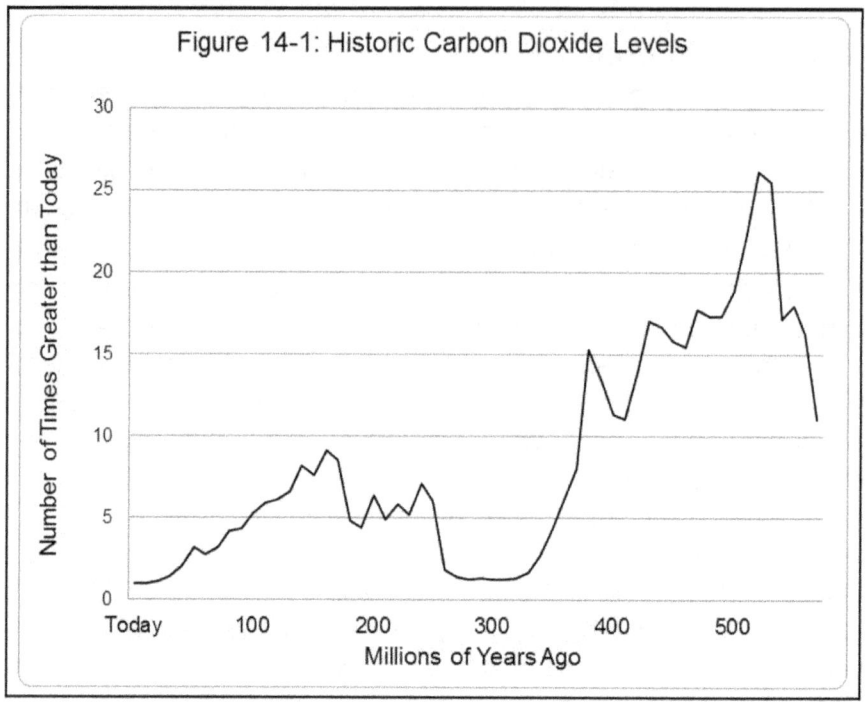

Figure 14-1: Historic Carbon Dioxide Levels

Figure 14-1 begs an important question. What is Earth's natural balance for CO_2? This figure also leads to an interesting, although crazy-sounding theory that says fossil fuels are renewable energy.

Fossil Fuels Are Renewable

This may be a crazy theory, but let's see where it goes. The energy we get from fossil fuels comes from organic material that died and decayed several million years ago. At some point in the food chain, this organic material came from plants. Plants grow through photosynthesis, absorbing CO_2, storing carbon (the stuff we like to burn), and releasing oxygen. It's all driven by the sun; so technically speaking, fossil fuels are stored solar energy. There's also another way to consider fossil fuels as renewable energy.

When it comes to CO_2, we can think of Earth and its atmosphere as a closed system. No CO_2 enters the system from outer space and no CO_2 escapes into outer space. All of the CO_2 we're emitting today has been in this closed system since our planet formed. When we burn fossil fuels today, we're literally returning CO_2 to the atmosphere where it was previously stored.

Imagine how lush plant life had to be to create the massive deposits of fossil fuel we're using today. Those plants apparently thrived in an atmosphere rich in CO_2. By comparison, today's atmosphere is starving for CO_2. Oceans, landmasses, and plants will eventually reabsorb the CO_2 we're emitting today. Those plants and animals will die and decay, and, under the right set of conditions, in a few million years they'll become fossil fuels once again.

Yes, this is a crazy theory because no one is going to wait a few million years for this unlikely source of renewable energy, but history indicates the process works.

There's another important consideration revealed in Figure 14-1. The Figure shows that since Earth's volcanic activity has slowed, Earth's *natural balance* has been draining CO_2 from the atmosphere. If left alone, would this natural balance eventually drain the atmosphere of all CO_2? If so, what would that mean for life on our planet? If this seems like another crazy theory, consider a recent satellite mission to Mars. On November 18, 2013, NASA launched a satellite bound for Mars with one goal in mind: To find out why Mars turned from a warm, moist planet to a dry, frigid world. Is it possible that over the past few million years, the natural balance on Mars drained all the CO_2 from its atmosphere, causing a disastrous change in climate?

Now we have two crazy theories. First, the theory that fossil fuels are renewable and then the theory that we may one day run out of atmospheric CO_2. At this point we might as well add a third crazy theory. To describe this last theory, I'll ask a question. When CO_2 levels were many times higher than today, did Earth experience a horrible climate of doom—as we're told to expect—or was the climate more that of a lush Garden of Eden? Given the dire warnings we've heard for more than twenty years, many are likely to see this as a ridiculous question. Is it?

Unstoppable Global Warming

Have you heard those claims indicating that a 15% increase above today's level of CO_2 will trigger unstoppable global warming?

It's a claim promoted by many, including nonprofit groups like *350.org*.[216] The group's name matches their goal of reducing CO_2 levels to their *natural balance* of 350 ppm. In their defense, members of this organization use the term *Tipping Point* rather than *Irreversible* or *Unstoppable*, but the media's interpretation has been clear. Reduce emissions of CO_2 or we'll trigger a global warming chain reaction we'll never be able to stop. History demonstrates the absurdity of this claim. CO_2 levels over 2,000% higher than today's levels didn't cause unstoppable global warming and couldn't prevent the major ice ages that subsequently occurred.

To repeat a previous question, is an atmosphere low in CO_2 better or worse than an atmosphere high in CO_2? The answer might surprise you.

The Most Intriguing Theory

The most intriguing theory I've uncovered in writing this book comes from a 1990s essay discussing the cause of ice ages. The essay is by Professor Fred Hoyle (June 24, 1915—August 20, 2001) and Professor Chandra Wickramasinghe.

Professor Hoyle is considered one of the great scientific thinkers of the twentieth century. He rose to become a world leader in astrophysics theory, was appointed to the illustrious Plumian Professor of Astronomy and Experimental Philosophy at Cambridge University, and became the founding director of the Institute of Theoretical Astronomy, an institute that quickly rose to become the world leader in theoretic astrophysics under Hoyle's leadership. Although several of his theories ran counter to mainstream scientific thinking, many of his controversial theories were subsequently validated.

His co-author, Professor Wickramasinghe, has written over 350 scientific papers on a variety of subjects. One of his most impressive discoveries was that interstellar dust contains organic matter, the building blocks of life. In 1973, he became the youngest department head ever appointed at University College, Cardiff when

---

216. *350.0rg*, "Science: The Basis of Climate Change Science," http://www.350.org/sites/all/files/science-factsheet-2010.pdf (Accessed 12-28-12).

he was named Professor and Head of the Department of Applied Mathematics and Mathematical Physics.

The essay in question makes it clear: water vapor is the unquestioned driver of the greenhouse effect, and the role of CO_2 is inconsequential when compared to other factors that drive climate change. Hoyle calls it "demented and dangerous" to pursue reductions in CO_2 to stop global warming. How can that be?

According to the essay, Earth's most recent one million years have been an ice age interrupted by short, highly beneficial warm periods, much like todays'. Hoyle and Wickramasinghe contend that adequate ocean evaporation is the primary process that prevents Earth from drifting back into ice age conditions. Water vapor serves as the conveyor belt transferring heat to the atmosphere. It also provides the atmosphere with the density, mass, and molecules required to absorb heat from the sun (think of the vacuum of outer space, which isn't warmed by the sun because it lacks the mass and density to absorb heat). According to the essay, a dry atmosphere absorbs and retains less heat from the sun, allowing the limited amount of remaining moisture to form ice crystals at high altitude. These ice crystals increase Earth's reflectivity (Earth's albedo), which then reflects needed solar heat back into space, locking earth into long-term ice age conditions.

The essay also contends that while ocean surface temperatures are warming, the ocean depths are cooling: an event that threatens to drive us back into a catastrophic ice age. Hoyle and Wickramasinghe warn that anything we can do to maintain robust ocean evaporation will benefit humanity, nature, and the planet in the long run. Essentially, his essay suggests that instead of trying to achieve the natural balance that causes repeated ice ages, we should concentrate on maintaining warm oceans at all cost. Hoyle has even suggested pumping water from the ocean depths in a controlled, long-term effort to warm the ocean and ensure that Earth maintains enough evaporation to avoid the next ice age.

Talk about your crazy theories! That's one for the record book, right? Perhaps, but you have to admit, it's an intriguing theory.

Today, we're told that Earth is entering the epoch of mass extinctions driven by human activity. The essay warns that the real suffering and mass extinctions are tied to ice ages, not warming.

Hoyle advised that the last thing we need to fear is global warming.

Argo

Argo is a new system of over 3,000 buoys scattered across the world's oceans. This network of buoys measures water chemistry and the heat content of the upper 2,000 feet of our oceans. Search the web for "Argo, cooling oceans" and you'll find a wide array of reports from the Argo project indicating our oceans have cooled since 2003. This is another seemingly critical fact the mainstream media has largely failed to mention. If this is true, ocean evaporation will soon begin to decline. Is it possible, Professors Hoyle and Wickramasinghe were correct when they advised that CO_2 is insignificant in terms of its impact on global warming?

More CarbonGate

According to Professor Zbigniew Jaworowski (now deceased), ice core records underestimate the level of pre-industrial CO_2 in the atmosphere. If that's true, todays' CO_2 levels might not be much higher than levels over the past several hundred thousand years.

Dr. Jaworowski published some 280 scientific papers and spent forty years involved in glacier studies in the Arctic, Antarctic, Alaska, Norway, the Alps, the Himalayas, and other locations. During his career, he worked with the Norwegian Polar Research Institute in Oslo, the Japanese National Institute of Polar Research in Tokyo, and helped the US EPA research the effects of industrial pollution on the global environment. Dr. Jaworowski began his study of glaciers in the early 1970s, long before talk of human-caused global warming. Dr. Jaworowski was not a member of the Flat Earth Society.

On March 19, 2004, Dr. Jaworowski wrote a paper for presentation to the US Senate Committee on Commerce, Science, and Transportation. The paper, titled "Climate Change: Incorrect Information on Pre-Industrial CO_2," details why and how CO_2 escapes from ice core samples during the extraction process—well before measurements can be taken. In other words, our measurements underestimate the actual value of pre-industrial CO_2 concentrations.

Believers object to this claim because Dr. Jaworowski was

challenging the *gold standard* for measuring CO_2 levels in ice cores. In other words, he was challenging the generally accepted "best practice" for measuring ice core CO_2. If we followed this same logic in medicine, we'd still be performing lobotomies. After all, the lobotomy procedure earned the Nobel Prize in Medicine; so at one time it was considered the *gold standard* for treating mental disorders.

Beyond flaws in the ice-core retrieval process, Dr. Jaworowski also indicated that CO_2 escapes before air is trapped in the ice. His contention was that chemical reactions continue to occur under the pressure of deep ice long before tiny pockets of ambient air are completely sealed-off. These on-going chemical reactions allow gases like CO_2 to escape before air bubbles are completely trapped. If these theories are correct, CO_2 levels before the Industrial Revolution were much higher than our scientists believe.

There's an even more troubling concern regarding today's understanding of past CO_2 levels. Dr. Jaworoski criticized what he called the IPCC's arbitrary rejection of ice-core samples that showed high levels of pre-industrial CO_2. In his words, this data was rejected because ". . . [It didn't] fit the pre-conceived idea on man-made global warming." Figure 14-2 on the next page is my rough recreation of Dr. Jaworoski's graph showing the biased selection of CO_2 measurements that were accepted by the IPCC.[217, 218]

---

217. Professor Z. Jaworowski, Prof T. V. Segalstad, and N. Ono, "Questioning the CO2 Ice Hockey Stick" (1992), The Science of the Total Environment, l14 (1992), 227-284; *Elsevier Science Publishers*, B.V., Amsterdam.
218. Z. Jaworowski, T.V. Segalstad, and N. Ono, "Do glaciers tell a true atmospheric CO2 story?"
http://www.greenworldtrust.org.uk/Science/Scientific/CO2-ice-HS.htm, and http://www.CO2web.info/stoten92.pdf (Accessed 6-10-12).

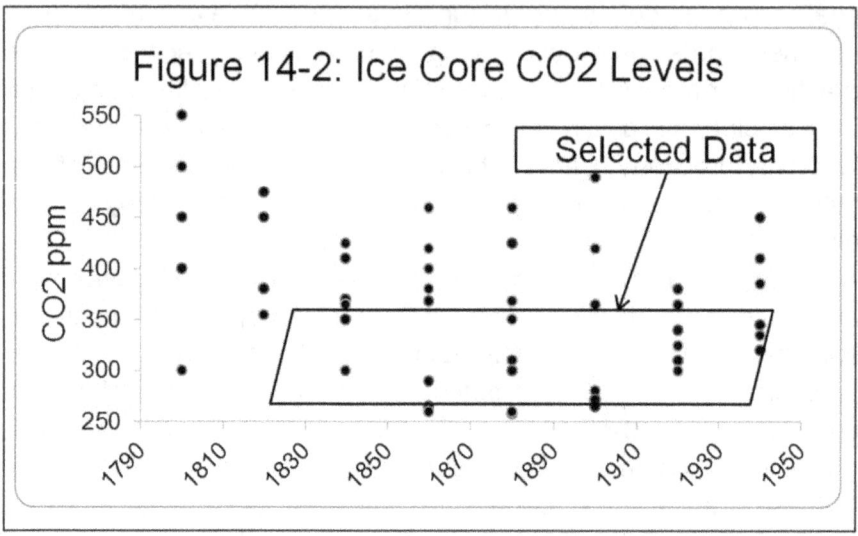

Figure 14-2: Ice Core CO2 Levels

Each dot in Figure 14-2 represents one of the historic CO_2 readings collected from ice cores. The measures shown in the boxed area labeled "Selected Data" are the only readings accepted by the IPCC. The other (higher) readings were rejected as "outliers." Rejecting higher readings and accepting only the lower measurements creates an obviously biased record of past CO_2 levels far below actual historic levels. According to Dr. Jaworoski, this level of "poor knowledge [incorrect data]" became widely publicized, widely accepted, and now serves as the foundation for all of the IPCC's work—and most, if not all, of our beliefs about rapidly increasing CO_2 levels.

It goes without saying that many disagree with Jaworoski's conclusions, but few critics can match his expertise or experience dealing with ice cores and CO_2 measures.

Diminished Impact of Increased CO_2?

If you search the web for "CO_2 impact is logarithmic," you'll find that scientists can't even agree on the impact of increasing levels of CO_2. Does doubling the amount of CO_2 double its heat trapping impact? Many argue that's not how it works. They say it's like applying sunscreen that blocks 50% of the sun's harmful rays. If you apply two layers, you don't block 100% of those harmful rays. The outer layer blocks 50% and the inner layer blocks 50% of the rays that pass through the outer layer. The extra layer is half as effective as the first.

If the impact of CO_2 is logarithmic, as many believe, that means increased levels of CO_2 have an exponentially decreasing impact on warming. You can draw your own conclusions from your web search, but your search will demonstrate at least one thing: Scientific consensus is the last thing that exists when it comes to CO_2's role in global warming.

High CO_2 during Ice Ages

It would certainly be a game changer if past ice ages occurred when CO_2 levels were many times higher than today. Surprisingly, numerous studies conclude that's exactly what Earth has experienced in the past—ice ages during periods of high CO_2.

A study by Dr. Daniel H. Rothman, professor of geophysics at the Massachusetts Institute of Technology (MIT), concluded that more than one ice age occurred when CO_2 levels were much higher than today's level. This study includes a graph of historic temperature and CO_2. The graph shows several past ice ages occurring during times when CO_2 levels were ten or more times greater than today.[219] This study, however, doesn't claim to refute the popular theory of today's warming. It simply concludes that CO_2 may not be a dominant driver of climate.

In case you're wondering, funding for Dr. Rothman's research group was provided by the Department of Energy; the Office of Basic Energy Sciences; the Geosciences Research Program; the National Science Foundation; Ocean Sciences, Emerging Topics in Biogeochemical Cycles; and the NASA Astrobiology Institute.[220]

Dr. Rothman's findings are also supported by *The New World Encyclopedia*, which discusses a major ice age that occurred

---

219. D.H. Rothman, "Atmospheric carbon dioxide levels for the last 500 million years" (January 17, 2002), Proceedings of the National Academy of Sciences USA 99: 4167-4171, http://www.pnas.org/content/99/7/4167.full (Accessed 7-3-10). Also available at
http://segovia.mit.edu/publications/2002/sretoc.pdf (Accessed 7-3-10).
220. "Financial Support," http://segovia.mit.edu/funding.html (Accessed 12-28-12).

roughly 450 million years ago. According to the encyclopedia, CO_2 levels during this ice age dropped from 7,000 ppm to around 4,000 ppm—CO_2 levels that were ten to seventeen times higher than today's level, yet the planet was deep in the grip of an ice age![221]

A study appearing in *Nature* recreated sea surface temperatures over the past 550 million years. When coupled with past records of CO_2, the results show several ice ages during periods of high CO_2. The study concluded that CO_2 could not have been the principal driver of historic climate changes.[222]

Believers challenge these studies.

The Believer's Rebuttal

Admittedly, there are many unknowns when trying to recreate 550 million years of climate history—Almost as many uncertainties as trying to predict Earth's future climate. Numerous researchers and websites use these uncertainties to attack the findings suggesting ice ages with high levels of CO_2.

The best rebuttal I've found comes from a website called *Skeptical Science*. One entry on this website is titled "Does high levels of CO_2 in the past contradict the warming effect of CO_2." This discussion warns that recreations of past CO_2 levels can include gaps in time spanning some ten million years. I couldn't validate this contention, but we'd all have to agree that a major drop in CO_2 could have occurred more than once in that length of time. If this website is correct, CO_2 levels during past ice ages may not have been as high as projected in the studies noted above.

It's an excellent point; however, this website goes on to discuss a major ice age that occurred when CO_2 levels were between 2,400 ppm and 9,000 ppm (six to over twenty-two times higher than

---

221. "Ordovician," *The New World Encyclopedia*, http://www.newworldencyclopedia.org/entry/Ordovician (Accessed 12-28-12).

222. Jan Veizer, Yves Godderis, and Louis M. Francois, "Evidence for decoupling of atmospheric CO2 and global climate during the Phanerozoic eon," *Nature*, Vol 408, December 7, 2000, 698. http://www.nature.com/nature/journal/v408/n6813/abs/408698a0.html (Accessed 12-29-12).

today's levels). The discussion reinforces the fact that our scientists believe that ice ages have occurred despite the presence of very high levels of CO_2. That's obviously not the conclusion reached by the website, however.

This rebuttal concludes that continental drift and the uplift of Earth's mountain chains were occurring at this same time. In theory, both of these events could have absorbed vast amounts of CO_2 leading to the ice age in question. The rebuttal's final conclusion also indicates that the sun was several percent dimmer than it is today. Therefore (according to this website), an ice age could have started with CO_2 levels around 3,000 ppm, or roughly seven times higher than today. If this rebuttal is true, perhaps the studies mentioned above are all wrong. Of course, in order to believe this one website, we'd have to dismiss a lot of scientific studies that show ice ages coinciding with high levels of greenhouse gases.

What to Believe

It's difficult to know what we can and can't believe about climate change, but one thing comes to mind: We can believe that we haven't been told the whole story—and that should scare us all.

Coral Bleaching and Ocean Acidification

As promised earlier in this chapter, let's turn our attention to the role CO_2 might play as the cause of coral bleaching and ocean acidification.

We've already seen that the UK High Court found that CO_2 couldn't be tied to coral bleaching. When bleaching occurs, coral turns white. Coral's color actually comes from the algae that take up residence in the coral's polyp tissues. As ocean conditions change, coral can be stressed, leading it to expel algae, causing bleaching. Interestingly, the color of algae is like the color of grass. It's created by photosynthesis. Algae grow by absorbing CO_2, storing the carbon, and releasing the oxygen. Therefore, it's been argued that increased levels of ocean CO_2 help algae grow, preventing coral bleaching.[223]

Unlike coral bleaching, ocean acidification is directly tied to CO_2. At least, that's the popular claim. Increased levels of CO_2 in the ocean are said to decrease the available calcium bicarbonate. Without calcium bicarbonate, oceans turn more acidic, shellfish like oysters and clams can't grow shells, and the skeletal growth of some species of coral slows. With little effort, we can find numerous studies and scientific papers indicating that ocean acidification is a growing problem driven by our use of fossil fuels.

We can also find studies that refute such claims. Perhaps the most enlightening discussion was presented by Dr. John T. Everett during a Senate hearing on *The Environmental and Economic Impacts of Ocean Acidification*. Dr. Everett was on the Board of Directors for NOAA's Climate Change Program before retiring from NOAA after a thirty-one year career. He was also a key contributor to many IPCC evaluations, including roles as lead author for the IPCC's fisheries studies, co-chair of the IPCC's Polar Regions studies, lead author for the IPCC's oceans studies, and co-chair of the IPCC's oceans and coastal zones studies.

In his Senate testimony, Dr. Everett indicated that in the same way land-based plants absorb CO_2 and release oxygen, ocean-based plants absorb CO_2 and release oxygen. When ocean CO_2 increases, ocean plants thrive providing added nutrients for ocean animals. When ocean animals thrive, they release increased amounts of calcium bicarbonate, the compound CO_2 is supposed to decrease.

Dr. Everett indicated that the IPCC climate modelers didn't take this increased production of calcium bicarbonate into account. He also indicated that the IPCC's latest study found no empirical evidence supporting the effects of acidification on marine biological systems. Further, he indicated that Earth has seen far greater increases in CO_2 in the past, and that today's ocean plants and animals evolved under conditions far more extreme than the worst-

---

223. Jason Buchheim, Director of Odyssey Expeditions, "Coral Reef Bleaching," Copyright 1998, Odyssey Expeditions- Marine Biology Learning Center Publications,
http://www.marinebiology.org/coralbleaching.htm (Accessed 1-23-12).

case scenario projected to occur due to human use of fossil fuels. As Dr. Everett put it, "If anything, the science indicates plants, at least, will be more successful, and since they are the bottom of the food chain; this cannot be totally bad."[224]

For the record, a friend informed me that Dr. Everett has been discredited by several websites. That much is to be expected for anyone challenging the environmental lobby. I visited a few of those websites and was not impressed by their arguments. One accused him of signing a Senate subcommittee petition that was sponsored by a senator who received campaign funds from Big Energy. This accusation seems puzzling. How was Dr. Everett influenced by campaign contributions received by a senator?

Another website discrediting Dr. Everett also claimed that the Little Ice Age was caused by the Bubonic Plague—go figure! Actually, I found several websites claiming that the Plague caused the Little Ice Age. The theory behind this claim says that one third of Europe's population was wiped out by the Plague. As a result, forest growth rebounded, soaking up so much CO_2 it plunged the planet into a few centuries of cold weather.

The choice is yours. Trust bloggers who believe the Bubonic Plague caused climate change, or trust one of the lead authors for the IPCC and his thirty-plus years of experience studying ocean systems. Before you make your choice, consider one question. If you choose the bloggers, what does that say about the credibility of the IPCC and their expert scientists?

Chapter Summary

Our perception of CO_2 as an environmental evil is incorrect. It's a perception that has been manufactured through creative marketing, sensational headlines, and missing facts. We haven't been told the whole story.

---

224. "Statement of Dr. John T. Everett, Joint Hearing on the EPA's Role in Protecting Ocean Health, May 11, 2010," 11, http://epw.senate.gov/public/index.cfm?FuseAction=Files.View&FileStore_i d=db302137-13f6-40cc-8968-3c9aac133b16 (Accessed 1-29-13).

Chapter 15
Meeting the Skeptics

"I think that those people [global warming Skeptics] are in such a tiny, tiny minority now with their point of view. They're almost like the ones who still believe that the moon landing was staged in a movie lot in Arizona and those who believe the Earth is flat."
—Al Gore (2008)[225]

Politicians have a rule: If you can't win based on your record (the facts), discredit your opponent. Al Gore is a politician.

Since Day One, Skeptics have been reviled and discredited. Before we even had a chance to hear the Skeptic's side of the story, we were told they were either puppets of Big Energy or crazed radicals who still think the Earth is flat. They've also been compared to the scientists who fought for the tobacco industry by claiming that smoking didn't cause lung cancer. Even President Obama joined the chorus, warning that we can't wait for a meeting of the Flat Earth Society before we start reducing CO_2 emissions. No wonder we don't trust the Skeptics!

We've also been warned that many leading Skeptics aren't even climatologists. For the record, most of the IPCC scientists aren't climatologists either. Most members of the IPCC never even studied the *cause* of global warming. Also for the record, Al Gore isn't a climatologist. He's no scientist, either.

Al Gore is a front man: a spokesperson whose tenure and success as the lead promoter of human-caused global warming has only been rivaled by the tenure and success of the Energizer Bunny.

---

225. Interview of Al Gore by Lesley Stahl, Produced by Richard Bonin and Karen Sughrue; "Al Gore's New Campaign," March 30, 2008; *60 Minutes*, http://www.cbsnews.com/stories/2008/03/27/60minutes/main3974389.shtml (Accessed 7-15-09).

That's actually a compliment because the Energizer Bunny is recognized as one of the most successful, long-lasting icons in the history of marketing, advertising, and promotion. Despite his obvious limitations, when Al Gore speaks, society listens—and they pay. His speaking fees start at $100,000 plus expenses.

We accuse the Skeptics of being paid to spread disinformation. Al Gore has large financial interests in companies that profit from low-carbon technologies. He's commonly referred to as the world's first global-warming billionaire. That's an exaggeration. When he left public office in 2001, his net worth was roughly $2 million. By late 2012, it was estimated to be around $100 million.[226] Hats off to Mr. Gore's shrewd investment strategies, but his adamant promotion of human-caused global warming while profiteering from green energy companies is at best self-serving and at worst a clear conflict of interest. We'd tar and feather a Skeptic with similar motives even if that Skeptic was a climatologist, yet we give Mr. Gore a free pass. It's another double standard we can't seem to avoid. It leaves us vulnerable to hearing only one side of the debate. It leaves us vulnerable to deception. Why do we listen to a man with such overwhelming profit motives and so little expertise on the subject matter?

Misconceptions

Before we meet a few of the world's leading Skeptics, there are a few misconceptions we need to clarify:

- Skeptics agree that global warming is real. They agree Earth is warming and the warming started at roughly the same time the Industrial Revolution was kicking into high gear.
- Skeptics don't oppose environmentalism. In fact, most of the skeptical scientists we'll meet have spent their entire careers fighting to support environmental causes. More

226. Carol D. Leonnig, "Al Gore has thrived as green-tech investor," Washington Post, October 10-2012
http://articles.washingtonpost.com/2012-10-10/politics/35502345_1_clean-energy-clean-tech-firms-al-gore (Accessed 3-13-13).

importantly, it's possible to be an adamant environmentalist and still oppose the belief that humans are the primary cause of global warming.

- Skeptics agree that human CO_2 emissions contribute to global warming, but they also believe CO_2 plays an insignificant role compared to other natural climate drivers.
- Skeptics know that severe weather, melting ice caps, rising oceans, and other warming-related events demonstrate the *effects* of global warming, but prove nothing about the *cause* of global warming.

Why Skeptics Exist

As previously asked, how could anyone, especially a scientist, dispute the overwhelming evidence collected by the IPCC? The last few chapters provide numerous reasons, but there's a more fundamental reason why many scientists remain Skeptics. The IPCC's *proof* ignores a few fundamental principles of science. First, science is based on observations. If those observations fail to support your theory, you have to find a new theory. Historic observations demonstrate that global warming has been a routine, recurring event driven by natural processes other than CO_2. History also shows that CO2 has always been a climate follower, not a climate driver. Ignoring these observations violates the fundamental definition of science. You may see this as splitting hairs, but if we're being completely honest and accurate, the IPCC cannot scientifically support its findings.

The second principle of science being ignored is called the Null Hypothesis. The Null Hypothesis essentially boils down to this: If IPCC scientists can disprove natural processes as the cause of today's warming, they could reasonably conclude that their theory is correct. The approach used by IPCC scientists has been the exact opposite. All of their efforts have focused on proving their own theory. As a result, they failed to rule out natural events as the cause of today's warming. It's a failure science should never accept.

For the scientific purist, the IPCC's failure to heed observed history and its failure to disprove natural causes are both fatal flaws. That may not change your belief in the IPCC or your view of

Skeptics, but it provides a valid reason why legitimate scientists remain skeptical. It also explains why the IPCC's formal assessment reports never claim to have proven the cause of today's warming.

Meeting Our First Skeptic

We'll begin our introduction of Skeptics with an individual who, like Al Gore, isn't a climatologist. Dr. Bjorn Lomborg is a long-time environmentalist, he is a past Director of Denmark's Environmental Assessment Institute, and as previously noted, he was a founding member of the prestigious environmental think tank, the Copenhagen Consensus Center. In 2004, *Time* magazine named him one of the world's 100 most influential people. In 2008, he was named one of the seventy-five most influential people of the twenty-first century by *Esquire* magazine, one of fifty people who could save the planet by *The Guardian*, and one of the top 100 public intellectuals by *Foreign Policy and Prospect* magazine. [227] He is also referred to as the *Skeptical Environmentalist*, the title of one of his books.

Dr. Lomborg doesn't attack the cause of global warming. Instead, he attacks the proposed cure: the IPCC's plan to save the planet. In his book, *Cool It*, he argues that the proposed reduction in greenhouse gases will have no meaningful impact on climate change while wasting trillions of dollars that could otherwise provide tremendous improvements in health and standards of living. He tells us that the plan to save the planet will do more harm than good.

Dr. Lomgborg indicates that a $140 tax on a ton of CO_2 would cut US carbon emissions in half, cost consumers $160 billion annually, and reduce the expected temperature in the year 2100 by a mere 0.2°F: a meaningless change. Similarly, Dr. Lomborg indicates that the EU's goal to reduce carbon emissions 20% by 2020 would delay the arrival of temperatures projected for 2100 until 2102. This two-year delay would cost consumers $90 billion annually.[228] This cost was later increased to $250 billion per year.[229]

---

227. http://www.lomborg.com/about/biography/ (Accessed December 24, 2009).
228. Dr. Bjorn Lomborg, "Cool It," Copyright 2007, 2008, *Vintage Books* a division of *Random House, Inc.*, xv.

Of course, these estimates of lower temperatures assume the IPCC is correct and CO_2 plays a meaningful role in climate change.

Lomborg warns that our focus on reducing emissions of CO_2 is misplaced. He insists that humanity would be better served by channeling funds into infrastructure improvements, including clean drinking water, sanitation, and irrigation to improve standards of living. He also warns that efforts to reduce CO_2 emissions will slow economic activity to the point where it will increase suffering caused by any future climate change.

Open Letter to the Canadian Prime Minister

The Kyoto Protocol (The Protocol) has now expired, but its significance lives on. The Protocol is the treaty the UN wanted all nations of the world to sign. Signing The Protocol committed developed nations to enforceable reductions in greenhouse gas emissions, but did not include enforceable reductions on undeveloped countries. On April 29, 1998, Canada became among the first developed countries to sign The Protocol. Formal ratification didn't come until four years later. As new rules and regulations began to take effect to meet the terms of The Protocol, many scientists came forward to object.

On April 6, 2006, sixty scientists from around the world sent an open letter to the Canadian Prime Minister, protesting the need for curbs on CO_2. The quotes listed below are taken from that open letter:

- Observational evidence [of our past and present climate] does not support today's computer climate models, so there is little reason to trust model predictions of the future. Yet this is precisely what the United Nations did in creating and promoting Kyoto and still does in the alarmist forecasts [of the dangers posed by climate change].
- Significant advances have been made since The Protocol was created, many of which are taking us away

---

229. Bjorn Lomborg, "Climate-Change Misdirection," *Wall Street Journal* (January 24, 2013), A15.

from a concern about increasing greenhouse gases. If, back in the mid-1990s, we knew what we know today about climate, Kyoto would almost certainly not exist, because we would have concluded it was not necessary.

- When the public comes to understand that there is no 'consensus' among climate scientists about the relative importance of the various causes of global climate change, the government will be in a far better position to develop plans that reflect reality and so benefit both the environment and the economy.

- [The expression] 'Climate change is real' is a meaningless phrase used repeatedly by activists to convince the public that a climate catastrophe is looming and humanity is the cause. . . . Allocating funds to 'stopping climate change' would be irrational.

- While the confident pronouncements of scientifically unqualified environmental groups may provide for sensational headlines, they are no basis for mature policy formulation.[230]

You can find the full letter and the names and credentials of the sixty scientists who signed this letter at the website referenced in the footnote at the end of the above quotes. These sixty scientists are emphatic on three points: 1) there is no scientific consensus regarding the cause of global warming; 2) computer models of our climate are not accurate and are of little use—other than creating alarmist headlines; and 3) efforts to reduce greenhouse gas emissions won't have a significant impact on Earth's climate.

Open Letter to the United Nations
A letter dated December 13, 2007 was sent to his Excellency Ban Ki-moon, Secretary-General of the United Nations. Copies were distributed to leaders of the countries participating in the IPCC. One hundred scientists from around the world, including many past

---

230 ."Open Kyoto to Debate,"
http://www.lavoisier.com.au/articles/greenhouse-science/climate-change/openletter2006-3.php (Accessed 7-1-10).

members of the IPCC, signed the letter. Key excerpts from the letter are shown below:

- The IPCC Summaries [assessment reports] . . . are the basis for most climate change policy formulation. Yet, these summaries are prepared by a relatively small core writing team with the final drafts approved line-by-line by government representatives. The great – majority [sic] of IPCC contributors and reviewers, and the tens of thousands of other scientists who are qualified to comment on these matters, are not involved in the preparation of these documents. The summaries therefore cannot properly be represented as a consensus view among experts.
- Attempts to cut [CO_2] emissions will slow development, the current UN approach of CO_2 reduction is likely to increase human suffering from future climate change rather than to decrease it.
- Recent observations of phenomena such as glacial retreats, sea-level rise and the migration of temperature-sensitive species are not evidence for abnormal climate change, for none of these changes has been shown to lie outside the bounds of known natural variability.
- Leading scientists, including some senior IPCC representatives, acknowledge that today's computer models cannot predict climate.
- In stark contrast to the often repeated assertion that the science of climate change is 'settled,' significant new peer-reviewed research has cast even more doubt on the hypothesis of dangerous human-caused global warming.[231]

---

231. "Open Letter to the Secretary-General of the United Nations" (December 13, 2007), http://scienceandpublicpolicy.org/images/stories/papers/reprint/UN_open_l etter.pdf (Accessed 9-7-2009).

Senate Minority Report

We can find more scientists disputing human-caused global warming by turning to a Senate Minority Report from the US Senate Committee on Environment and Public Works (EPW). This EPW report includes comments, submissions, and papers from over 700 scientists objecting to the work and conclusions of the IPCC. It's worth noting that in 2007 only 400 scientists were listed in the EPW report. The number of skeptical scientists appears to be growing, not shrinking.[232]

It goes without saying that Believers discredit this EPW website. Believers claim the website has its roots in conservative ideologies and interests like Big Energy. That's normally more than enough to discredit the report, but before we summarily disregard the thoughts of 700 scientists, let's first see what they are saying. A few of the more revealing quotes from this website are listed below:

- "I have found examples of a Summary [an IPCC assessment report] saying precisely the opposite of what the scientists said." —Dr. Philip Lloyd, IPCC lead coordinating author, South Africa.
- "[Al] Gore prompted me to start delving into the science again and I quickly found myself solidly in the skeptic camp."—Meteorologist Hajo Smit, former member of the Dutch IPCC committee.
- "Even doubling or tripling the amount of carbon dioxide will virtually have little impact."—Professor Geoffrey G. Duffy, New Zealand.
- Human-caused global warming is the " . . . worst scientific scandal in the history [of humanity]. . . . When people come to know what the truth is, they will feel deceived by science and scientists."—Dr. Kiminori Itoh,

232. U. S. Senate Minority Report: "More Than 700 International Scientists Dissent Over Man-Made Global Warming Claims. Scientists Continue to Debunk 'Consensus' in 2008 & 2009" (updated March 16, 2009), http://www.epw.senate.gov/public/index.cfm?FuseAction=Minority.Blogs&ContentRecord_id=2674E64F-802A-23AD-490B-BD9FAF4DCDB7 (Accessed 7-10-10).

environmental physical chemist and former member of the Japanese IPCC committee.

- "It is a blatant lie put forth in the media that makes it seem there is only a fringe of scientists who don't buy into anthropogenic [human-caused] global warming."— Stanley B. Goldenberg of the Hurricane Research Division of the National Oceanic and Atmospheric Administration (NOAA).

- "The IPCC has actually become a closed circuit; it doesn't listen to others. It doesn't have open minds. . . . I am really amazed that the Nobel Peace Prize has been given on scientifically incorrect conclusions. . . ."—India geologist Dr. Arun D. Ahluwalia from Punjab University and board member of the UN-supported International Year of the Planet.

- "The [global warming] scaremongering has its justification in the fact that it is something that generates funds."—Paleontologist Dr. Eduardo Tonni of the Committee for Scientific Research in Buenos Aires and head of the Paleontology Department at the University of La Plata.

- "The Kyoto theorists have put the cart before the horse. It is global warming that triggers higher levels of carbon dioxide in the atmosphere, not the other way round. . . . A large number of critical documents submitted at the 1995 UN conference in Madrid vanished without a trace. As a result, the discussion was one-sided and heavily biased, and the UN declared global warming to be a scientific fact."—Andrei Kapitsa, Russian geographer and Antarctic ice core researcher.[233]

One of the links you'll find on the EPW website is a public letter detailing the experience of Dr. Christopher W. Landsea of NOAA's

---

233. Ibid.

National Hurricane Center.[234] He was both an author and a reviewer for the IPCC's second assessment report (1995) and its third report (2001). He resigned from the fourth report, claiming the UN was playing politics with hurricane science. His public letter included two telling statements:

- I am withdrawing [from the IPCC] because I have come to view the part of the IPCC to which my expertise is relevant as having become politicized. In addition, when I have raised my concerns to the IPCC leadership, their response was simply to dismiss my concerns.
- I personally cannot in good faith continue to contribute to a process that I view as both being motivated by pre-conceived agendas and being scientifically unsound.

Two More Skeptics

I'll mention two additional scientists I consider Skeptics. These two Skeptics are so reviled by Believers, one Believer told me he could "filet" their reputations in a matter of minutes with a simple web search. I have no doubt that's true. The character of anyone challenging the popular theory of global warming is immediately attacked. If you read these attacks, you'll be convinced of one of two things. These Skeptics are either truly evil people intent on destroying the planet to line their pocketbooks, or the personal attacks are so outlandish and inconsequential, they do more to discredit the Believers than the Skeptics.

At the risk of having this chapter discredited because I dared to present the views of these two Skeptics, I thought it was worth hearing what they had to say.

The first scientist is Dr. John Christy from the Atmospheric Science Department and Director of the Earth System Science Center at the University of Alabama, Huntsville. Dr. Christy was one

---

234. "U.S. Senate Minority Report," (December 11, 2008), 141, http://epw.senate.gov/public/index.cfm?FuseAction=Files.View&FileStore_i d=83947f5d-d84a-4a84-ad5d-6e2d71db52d9&CFID=50946553&CFTOKEN=51662232 (Accessed 7-10-10).

of the lead authors of the IPCC's Third Assessment Report (2001) and was also on the panel developing the American Geophysical Union's (AGU) landmark 2003 statement on climate change. That statement essentially reaffirmed the role of human activities as the cause of climate change. It appears Dr. Christy's beliefs have changed since that 2003 statement.

Dr. Christy indicates that scientists who signed the IPCC reports were only signing their individual contributions and were never asked to sign the report's final conclusions. He contends the UN edited the scientists' contributions—after scientists signed their work. This editing is part of the highly regarded, highly praised peer-review process that allegedly ensures scientific consensus. In reality, it appears this so-called *peer-review* often reversed the conclusions of the participating scientists.[235]

Another Skeptic, Professor Paul Reiter of the Pasteur Institute, Paris and contributor to the IPCC's third assessment report ultimately resigned from the IPCC over this same issue. In a memorandum submitted to the United Kingdom's House of Lords on March 31, 2005, Dr. Reiter reviewed his IPCC experience. Sections of his memorandum are presented here:

- Not one of the lead authors [in Dr. Reiter's area of research] had ever written a research paper on the subject! Moreover, two of the authors, both physicians, had spent their entire career as environmental activists.
- The amateurish text of the chapter reflected the limited knowledge of the 22 authors.
- The treatment of this issue by the IPCC was ill-informed, biased, and scientifically unacceptable.
- The third assessment report listed more than 65 lead authors [in Dr. Reiter's area of expertise], only one of which . . . was an established authority on the health chapter.
- My colleague and I repeatedly found ourselves at

---

235. John Stossel, "Is there really a global warming consensus?" *ABC 20/20* (October 22, 2007), http://www.youtube.com/watch?v=BZcp_wcDXec (Accessed 7-11-2009).

loggerheads with persons who insisted on making authoritative pronouncements, although they had little or no knowledge of our specialty.[236]

According to Dr. Reiter, the IPCC's conclusions were unscientific and biased, the IPCC scientists aren't necessarily experts in the areas they studied, and many were environmental activists. Dr. Reiter reminds us that the IPCC is a "governmental" panel, meaning politicians appointed those serving on the panel. Dr. Reiter has called the IPCC's conclusions "unmitigated rubbish."[237]

Perhaps Dr. Reiter has sold out to the Dark Side (Big Energy), perhaps he is simply a disgruntled ex-IPCC veteran, or perhaps his experiences and concerns are valid and symbolic of a much wider problem within the IPCC?

What to Believe?

Should we trust the Skeptics or ignore them? Is that even the right question? A better question might ask why we haven't heard the Skeptics' side of the debate. Is the mainstream media protecting us by blocking propaganda from a well-financed disinformation network, or is the media dictating what they want us to believe?

If a well-financed disinformation network is spewing false information, perhaps that network isn't comprised of the Skeptics. Perhaps it's the media, the UN, Hollywood, and the world's first global-warming billionaire.

---

236. Dr. Paul Reiter, "Memorandum: The IPCC and technical information, example: Impacts on Human Health"(March 31, 2005),
http://www.publications.parliament.uk/pa/ld200506/ldselect/ldeconaf/12/12 we21.htm (Accessed 6-20-10).
237. John Stossel, "Is there really a global warming consensus?" *ABC 20/20* (October 22, 2007),
http://www.youtube.com/watch?v=BZcp_wcDXec (Accessed 7-11-2009).

Chapter 16
Flares, Tilt, Wobble, and Whirl

"What we've got here is failure to communicate."
—*Cool Hand Luke*

We believe CO_2 drives climate change because no one has given us anything else to blame. The media, the UN, Hollywood, and the environmental lobby have emphasized only one known climate driver: our greedy appetite for fossil fuels. Since we have no other suspects in our quest to name a guilty party, it's easy to believe that CO_2 must be the lone cause.

Unfortunately, it's a situation that has occurred in real life criminal trials. Detectives and prosecutors focus their efforts on the prime suspect, the circumstantial evidence supports their position, they present only one side of the story, and they build a seemingly airtight case against their lone suspect. Based on the only evidence presented, the jury subsequently (and understandably) finds the suspect guilty—only to have the suspect exonerated years later when DNA proves his or her innocence.

When it comes to climate change, there's no shortage of prosecuting attorneys who desperately want us to believe that CO_2 is guilty. We'll explore those prosecutors and their motives in detail in later chapters, but for now, let's focus on potential suspects.

Why It *IS* the Sun

Let's start with the IPCC admission that low solar activity cools the planet. That much validated by *Environmental History Resources*, which indicates that one of the coldest portions of the Little Ice Age occurred during a period corresponding to the Maunder Minimum: a period of surprisingly low sunspot activity as recorded by one of humanity's earliest astronomers, Galileo.[238] The Astronomical Society of the Pacific also notes a different colder period within the Little Ice Age, which coincided with the Dalton Minimum—another period of low sunspot activity occurring roughly between 1795 and the 1820s.[239] If low solar irradiance cools the planet, doesn't it stand to reason that increased solar irradiance would warm the planet?

In January 2013, NASA summarized a National Research Council (NRC) study discussing new and surprising findings related to the impacts of changing solar irradiance. The NRC found that a change as small 0.1% in solar irradiance produces a ten-fold change in extreme ultraviolet energy (EUV). According to the NRC, changing EUVs have a major impact on the chemistry and temperature of the upper atmosphere, which translates into larger changes in climate than previously believed. It reduces protective ozone, allowing more UV radiation to reach Earth's surface. It causes significant changes in wind patterns from the upper atmosphere all the way down to Earth's surface. It also impacts a number of other natural climate drivers.[240]

Numerous studies of solar activity clearly show that changes in the sun have controlled Earth's climate changes over the past 300 years. According to NASA's Cosmos series, ". . . the Sun's changing brightness dominated our climate for two centuries, from 1600 to 1800. . . . the Sun [also] noticeably warmed the climate for another century . . ." beginning in roughly 1870.[241] A report from the

---

238. "The Little Ice Age, Circa. 1300–1870," Environmental History Resources,   http://www.eh-resources.org/timeline/timeline_lia.html#maunder (Accessed 6-13-10).
239. Willie Soon and Stephen H. Yaskell, "Year Without A Summer," *Astronomical Society of the Pacific* (May/June 2003), http://www.astrosociety.org/pubs/mercury/32_03/summer.html (Accessed 6-13-10).
240. Dr. Tony Phillips, "Solar Variability and Terrestrial Climate," *Science@NASA*, http://science.nasa.gov/science-news/science-at-nasa/2013/08jan_sunclimate/ (Accessed 1-14-14).
241. NASA's Cosmos, "4. Sun-Earth Connection," http://ase.tufts.edu/cosmos/view_chapter.asp?id=34&page=4 (Accessed 12-14-13).

Goddard Space Flight Center titled "NASA Study Finds Increasing Solar Trend Can Change Climate" indicates the overall intensity of the sun has steadily increased every decade since the 1970s. This report also states, "Historical records of solar activity indicate that solar radiation has been increasing since the late nineteenth century."[242] This timing coincides with the start of the Industrial Revolution and, not surprisingly, the end of the Little Ice Age.

Skeptics also point out that Earth warmed faster prior to the 1930s than it's warming today. Earth also cooled from the 1940s through the early 1970s, before a return to warming between roughly 1975 and at least 1998. These changes don't parallel increases in human CO_2 emissions, which began skyrocketing in the 1950s and 1960s, but they do match changes in the sun.

The relationship between solar activity and climate change is so powerful, the IPCC and environmentalists are changing their entire story about human-caused global warming. It's a subtle change that may have gone unnoticed, but the change is profound.

In case you've missed it, the IPCC is no longer telling us that human-caused global warming started with the Industrial Revolution. In their fifth assessment report, the expert opinion of IPCC scientists indicated that " . . . human influence on climate caused more than half of the observed increase in global average surface temperature **from 1951–2010**" [Emphasis Added]. Apparently, the IPCC couldn't explain why Earth warmed faster prior to the 1930s than it's warming today, so they had to change the date when human CO_2 emissions began influencing climate.

What happened to the twenty-plus years of *undeniable proof* that human activities started controlling climate with the start of the Industrial Revolution? Additionally, if we began controlling climate in 1951, why did cooling persist until the mid-1970s? Is seems we can only accuse humans of causing global warming if we pick a cold period in history to begin tracking temperatures. The first starting point was the Little Ice Age. Now it's the cool period between the

---

242. "NASA Study Finds Increasing Solar Trend Can Change Climate" (March 20, 2003),
http://www.nasa.gov/centers/goddard/news/topstory/2003/0313irradiance.html (Accessed 12-21-11).

1940s and mid-1970s. Couldn't we accuse the IPCC of cherry-picking their starting dates just as the Skeptics were accused of cherry-picking 1998 as the start of recent global cooling?

Solar Winds

Changing solar activity includes more than changing heat output. As we've seen, it includes changes in ultraviolet radiation. It also includes changes in solar winds and changes in the visible and infrared spectrum. These changes have far-reaching climate impacts: some are well understood; others are not.

Solar winds are fast-moving charged particles streaming away from the sun. These particles can damage chromosomes and cause cancer. Fortunately, Earth's magnetic field helps protect us from these particles. Unfortunately, Earth's magnetic field doesn't always prevent solar winds from wreaking havoc. Solar winds caused the widespread 1989 electric blackout in Canada. Solar winds may also have been responsible for relocating Earth's magnetic North sometime between 900 and 1500 B.C.—a time corresponding to the Medieval Climate Optimum and the Viking settlements in Iceland.

Studies indicate that sunspots (and the associated solar winds) have doubled over the past 100 years.[243] Skeptics argue it's no coincidence that Earth has warmed during this same period. In fact, the Schroeder Institute for Research in Cycles of Solar Activity indicates that, "Near-Earth variations in the solar wind . . . since 1868, are closely correlated with global temperature."[244]

---

243. David Noss, "The Role of Sunspots and Solar Winds in Climate Change," *Scientific American*, July 22, 2009, http://www.scientificamerican.com/article.cfm?id=sun-spots-and-climate-change (Accessed 11-19-11).
244. Theodor Landscheidt, "Solar Wind Near Earth: Indicator of Variation in Global Temperature," Schroeter Institute for Research in Cycles of Solar Activity. Proceedings of 1st Solar & Space Weather Euroconference, 'The Solar Cycle and Terrestrial Climate,' Santa Cruz de Tenerife, Tenerife, Spain, September 2000 (ESA SP-463, December 2000), Abstract, http://www.mitosyfraudes.org/Calen/SolarWind.html (Accessed 11-19-11).

More Solar Cycles

Beyond the sun's eleven-year sunspot cycle, the sun also has repeating cycles lasting 22 years, 1,500 years, 11,000 years, and 100,000 years. In a book by S. Fred Singer and Dennis T. Avery titled *Unstoppable Global Warming Every 1,500 Years*, we're warned that the sun is currently at or near the end of its highest output over its 1,500-year cycle. Is warming the only climate change we'll experience over the coming centuries?

CERN

The climatic impact of cosmic rays has received little media attention, but recent studies by the European Organization for Nuclear Research (CERN, 2011) suggest that changes in both cosmic energy and solar winds affect cloud formation, which plays a larger role in climate change than previously suspected.[245]

Believers attack the CERN results, indicating that the findings are preliminary and based on laboratory experiments. They claim such experiments are of little significance compared to the overwhelming evidence from the IPCC. This is rapidly becoming the only argument Believers can use: Humans cause climate change because the IPCC says we cause climate change.

Solar Inertial Motion

The sun also changes position inside our solar system every 179 years. This is caused by changing gravitational pull generated by regular changes in the alignment of planets. The sun's movement over this cycle includes two types of movement: chaotic and stable. During periods of chaotic movement, solar minimums occur. Solar minimums are periods of low solar flares and low energy output. Four of these solar minimums occurred during the Little Ice Age. During periods of stable movement, solar maximums occur. Solar maximums are periods of increased solar energy that

---

245. "CERN's CLOUD experiment provides unprecedented insight into cloud formation," press release (August 25, 2011), http://press.web.cern.ch/press-releases/2011/08/cerns-cloud-experiment-provides-unprecedented-insight-cloud-formation (Accessed 11-26-11).

act to warm our planet. [246]

In 1987, a new cycle of changing solar position was also discovered. The media said little about this discovery. The IPCC ignored it. This cycle of changing solar position occurs roughly every 2,400 years. NASA calls this change in solar position the Charvatova Cycle in honor of Professor Ivanka Charvatova who discovered it. The cause is not yet known, but within this cycle, there appears to be a 370-year stable period of movement that results in a long-term solar maximum and global warming. When asked how the IPCC uses this new information in their modeling, Professor Charvatova indicated, "They don't. They pretended it didn't happen."[247]

Ocean Oscillations

There are repeating exchanges of cold water circulating from ocean depths to the surface. These exchanges are driven by changing temperature, changing pressure, and the subsequent changing wind patterns they create. The results include climate events including El Nino (warm ocean surface temperatures) and La Nina (cool ocean surface temperatures). A growing body of science identifies ocean oscillations as a critical driver of climate. The warming and cooling influences of these oscillations have been linked to historic 30-year cycles including warming during the periods of roughly 1850-1880, 1910-1940, and 1970-2000. This 30-year cycle is also linked to cooling including the period of the 1940s to the 1970s. Based on the influence of these cycles, many scientists predict we've entered a period of cooling that will last until 2030.[248]

---

246. Vítězslav Kremlík,"Interview: Is climate change caused by solar inertial motion?" (May 2011), The Geophysical Institute of the Czech Academy of Sciences, http://www.klimaskeptik.cz/news/interview-with-dr-ivanka-charvatova-csc-from-gfu/ (Accessed 6-30-13).
247. Ibid.
248. Nicola Scafetta, "Solar and planetary oscillation control on climate change: hind-cast, forecast and a comparison with the CMIP5 GCMs," July 16, 2013, 1Active Cavity Radiometer Irradiance Monitor (ACRIM) Lab, Coronado, CA 92118, USA, & Duke University, Durham, NC 27708, USA, http://arxiv.org/pdf/1307.3706.pdf (Accessed 12-14-13).

Are ocean oscillations one of the suspects the prosecuting attorney failed to identify in their quest to find those guilty of causing climate change?

Earth's Tilt, Wobble, and Whirl

We're all familiar with Earth's changing *tilt*, which causes our annual change of seasons. In addition, Earth has a wobble that occurs on a repeating cycle every 41,000 years. This wobble causes Earth's tilt (its alignment with the sun) to vary from roughly 22 degrees to 25 degrees. When the wobble causes a larger tilt, the change of seasons is more severe: summers are warmer and winters are colder. When the wobble causes a smaller tilt, NOAA's National Climatic Data Center indicates, "It is the cool summers which are thought to allow snow and ice to last from year to year in high latitudes, eventually building up into massive ice sheets."[249]

Earth's orbit around the sun—let's call it Earth's *whirl*—is not circular. It's oblong and it has a repeating 22,000-year cycle. The oblong whirl places Earth closer to the sun at different times of the year. For example, Earth is currently closest to the sun in January of each year, but 11,000-years ago Earth was closest to the sun in July. Today's position creates milder winters in the Northern Hemisphere and warmer summers in the Southern Hemisphere: two events that serve to increase annual average global temperatures. Makes you wonder, doesn't it?

Earth's whirl also includes cycles of 100,000 years and 400,000 years. According to NOAA, ". . . [It is the combination of orbital cycles that] affect the relative severity of summer and winter, and are thought to control the growth and retreat of ice sheets."

When we combine Earth's repeating cycles of tilt, wobble, and whirl with repeating cycles of solar activity, solar positioning, and ocean oscillations, we get vastly changing climates that are driven by natural events other than CO_2. Also, history has shown that CO_2 is a climate follower, not a climate driver. Are we now supposed to ignore history because the IPCC says we can? Are we certain CO_2

---

249. National Oceanic and Atmospheric Administration; National Climatic Data Center, "Paleoclimatology, Astronomical Theory of Climate Change," http://www.ncdc.noaa.gov/paleo/milankovitch.html (Accessed 4-3-10).

is the only suspect we should consider in our search for suspects causing today's climate change?

No one disputes the fact that increased CO_2 contributes to global warming. The dispute is whether or not that contribution plays a major role or a minor role. We've seen a NASA report indicating the sun's radiation has been increasing since roughly the start of the Industrial Revolution. We've seen evidence that low solar activity coincides with colder climates, ocean oscillations coincide with climatic warming and cooling, and ice ages are thought to have occurred during periods of very high atmospheric CO_2 levels. We also know that Earth's tilt, wobble, and whirl drive major changes in our climate. Despite all of this evidence, the IPCC insists we can trust that CO_2 is the lone guilty party causing global warming. Can we trust the IPCC?

The Bottom Line

We blame CO_2 because that's the only suspect we've been given. The prosecution has built an airtight case against CO_2, but the case isn't closed. The hidden facts reveal details that point to many guilty parties. Believe what you want regarding the cause of global warming, but you can't deny that critical facts have been hidden from your view.

Someone is trying very hard to prejudice the jury—we have no idea how hard they're actually trying.

Chapter 17
Marketing Trumps the Truth

Human-caused global warming is like a brand that can be sold.
This is the course we must follow to change the public's beliefs and
behaviors.
—Paraphrased: The Institute for Public Policy Research (IPPR) [250]

We know that Big Energy and Big Business hire marketing consultants not only to sell their products, but also to improve their public image and sway the opinions of voters. Speaking more frankly, we'd say Big Energy hires marketing gurus to sell us down the river.

Therefore, it's only fair that environmental organizations hire their own marketing experts to help promote their side of key issues, especially an issue as important as global warming. That being the case, how does one *market* global warming and climate-friendly behavior? Two reports provide an insider's view of the marketing recommendations given to environmental groups promoting the need for climate action. One of those reports, prepared by the Institute for Public Policy Research (IPPR, UK), is titled "Warm Words: How are we telling the climate story and can we tell it better." EcoAmerica, a market-consulting firm serving environmental organizations, prepared the other report. The EcoAmerica report is titled "Climate and Energy Truths, Our Common Future: Making the

---

250. Nat Segnit, Gill Ereaut, "Warm Words: How are we telling the climate story and can we tell it better?" (August 3, 2006), 8-9, Institute for Public Policy Research,
http://www.ippr.org/images/media/files/publication/2011/05/warm_words_1 529.pdf (Accessed 5-13-13).

Necessary Connections."[251]

A few of the recommendations from these reports are paraphrased in the list shown below:

- Our task is not to persuade by rational arguments, but to work in more clever ways.
- Always insist that the argument has been won.
- Tell your audience that the only remaining Skeptics are paid experts.
- Treat the facts as such a given, ". . . they [the facts] need not be spoken."
- Myth can be a powerful tool of persuasion.
- Talk about human values including clean air, energy independence, and creating new jobs, but don't discuss specifics because that can level the playing field.
- Don't talk about cap-and-trade, but talk about a "pollution reduction fund" or "cap and cash back."
- Avoid discussing the costs.
- Don't differentiate various forms of pollution [link CO_2 to all forms of pollution].
- Alarmism can be a good thing if your audience knows solutions are available. Sell the concept that tiny actions can save the planet.

You likely recognize several of these recommendations: "The debate is over," "we need to create green jobs," and "the Skeptics all work for Big Energy." When repeated often enough, these phrases become so taken for granted, the facts "need not be spoken." You have to admit, this is sound marketing advice, but it doesn't exactly lead to presenting the fundamental facts. It's quite the opposite. Environmentalists are instructed to ignore the facts, the costs, and rational argument, and instead use creative word

---

251. EcoAmerica, "Climate and Energy Truths, Our Common Future: Making the Necessary Connections," (April 2009), http://ecoamerica.org/wp-content/uploads/2013/02/Climate_Energy_Truths.pdf (Accessed 6-19-13).

choices that target the value systems of voters.

This is the marketing we expect from Big Business, not our trusted environmental organizations. Perhaps we should give environmentalists a free pass? After all, they're simply leveling the playing field, aren't they?

Leveling the Playing Field

Many would argue that EcoAmerica and the IPPR are leveling the playing field against the big-money marketing campaigns of Big Energy. That, of course, assumes the playing field was level in the first place.

You can judge for yourself, but isn't it true that the playing field of public opinion has been slanted in favor of environmentalism since the 1962 release of *Silent Spring*? We're constantly hearing about disinformation from industry, but it's clear we're not exactly getting the unbiased truth from the environmental lobby. If you think industry has the upper hand or if you believe the playing field is level, consider one question. Who do you trust and believe: Big Energy or the environmental lobby?

The honest answer is universally the same. Societies the world over mistrust industry and embrace the words of the environmental lobby. The playing field is far from level, and it hasn't been level for decades.

We saw how the public reacted to Dr. Wakefield's study linking childhood vaccines to autism. When some twenty subsequent government-sponsored studies found no link, many accused those studies of presenting disinformation to protect industry. Even the GMC was accused of providing cover for the vaccine industry. If we believe industry money slants the playing field in favor of the Skeptics, perhaps we should consider the hundreds of billions of dollars governments are spending on global warming research. In all honesty, which side of this debate is being influenced by Big Money?

Promoting environmental issues with a philosophy that believes the facts need not be spoken isn't leveling the playing field; it's stacking the deck and paying-off the referees to ensure the desired outcome. It's an affront to basic honesty, and yes, it should outrage us all, but it doesn't.

We're not outraged by the environmental lobby. We're outraged by Big Business and those darned Skeptics. Is our outrage misplaced? The groups we most trust—our environmental watchdogs and many other public interest groups—are the ones who avoid the fundamentals, skip the details, and apply double standards whenever necessary.

With global warming, we're dealing with the alleged Mother of all crises and the most expensive public policy in human history—and we're moving forward without hearing the whole story. It doesn't matter how much we hate Big Business, how strongly we support environmentalism, or how deeply we support reductions in greenhouse gas emissions, we can't, or at least we shouldn't, accept such intentional deception.

To call it an outrage is an understatement, yet we naively continue to trust. Such is the power of the environmental lobby and the influence of the mainstream media.

We've been deceived and the deceptions are intentional. The only remaining question is why. Why are so many organizations trying so hard to convince us that humans cause today's warming?

Chapter 18
Means, Motive, and Opportunity

"You never want to let a good crisis to go to waste."
—Rahm Emanuel, former White House chief of staff

If we return to the issue of a criminal trial accusing CO_2 of murdering our climate, the prosecution is using a philosophy that says myths can sway the jury and the facts "need not be spoken." Perhaps the trial we should be pursuing is one accusing the prosecution of fraud.

If we were on the jury in a case about prosecutorial fraud, the first thing we'd want to know is motive. Why would the prosecuting attorneys (the IPCC and others) want to deceive the entire world about global warming? The next thing we'd want to know is means and opportunity. It goes beyond reason to believe someone could deceive the entire planet; therefore, it's difficult to understand how anyone could have the means and the opportunity to achieve such an unlikely feat.

In reality, such a grand deception is much easier to achieve than we'd like to believe. We'll explain why in a few pages, but let's first focus on motive.

A Good Crisis

If you want to induce mass behavior change, a good old-fashioned crisis is the best way to accomplish your goal. World War II was a classic example. The vast majority of Americans wanted to stay out of World War II, but that opinion changed with the Japanese surprise attack on Pearl Harbor. The attack provided the crisis that instantly ignited American rage and mobilized the entire population to go to war.

Today, we're at war against fossil fuels. We've been in this war for decades, but previous crises weren't enough to spur the masses to change our wasteful ways. Acid rain, premature death, energy shortages, and respiratory disease didn't change our behavior. A new crisis was needed to spur us to action. Global warming represents the perfect crisis: the end of an inhabitable planet. For anyone wanting to change our energy and environmental behavior, blaming CO_2 for global warming (and numerous other evils) is the dream crisis, come true. It's the crisis no one wants to let "go to waste."

UN Motives

The UN is leading the charge to promote human activities as the primary cause of global warming. They keep global warming in the headlines by routinely staging worldwide Climate Conferences and by publishing new climate reports roughly once every five years. If global warming isn't a real crisis, why would the UN go to so much trouble?

Also, the UN is generally held in high regard. We perceive the UN as a pure and just organization as well as a peace-loving protector of all nations and peoples. It's difficult to image the UN pursuing a philosophy that believes the facts "need not be spoken."

On the other hand, the UN is, by definition, the most political organization on the planet. The UN is all politics, all the time—and the majority of its members are undeveloped nations. Could that somehow provide a motive for their promotion of human activities as the cause of global warming?

To find out, we can examine how the UN defines itself. What is its mission statement? What are its founding principles and its primary objectives? If you search the web for "United Nations Objectives," the bullet points below are among the first things you'll find (accessed September 10, 2009):

B. Objectives [of the UN]
- Strengthening international cooperation for develop-ment.
- Implementing all international agreements and com-mitments for development.
- The new opportunities, challenges, and risks opened by the [world's economic] globalization . . . heighten the

need for strengthened international cooperation. A strong political will is essential to sustain such cooperation. Through this Agenda, we renew our commitment and seek to impart new vigor to a global partnership for development.

One of the UN's primary objectives is to speed development of Third World countries through partnerships with developed nations. It's a noble cause and one we should find easy to support, but how can global warming help achieve this goal?

The answer lies in understanding two fundamental issues. The first issue is the vital role that low-cost energy plays in spurring economic growth, improving standards of living, and improving human health. We've already seen that The UN and the World Health Organization specifically cite access to affordable energy as the key for achieving these goals. Korea validates the importance of a strong economy and the link between low-cost energy, increased economic activity, and improved health and increased standards of living.

So how does global warming help provide low-cost energy to Third World nations? The answer lies in the second fundamental issue we need to understand: the terms of the Kyoto Protocol.

The Kyoto Protocol

As previously mentioned, the now expired Kyoto Protocol (The Protocol) was an international treaty that placed limits on greenhouse gas emissions from developed nations, but did not require reductions from Third World countries—even though those nations were already emitting 20% more greenhouse gases than developed nations. In other words, The Protocol addressed less than half of CO_2's 0.5% contribution to the greenhouse effect. Most would reasonably see this as a fatal flaw, and it revealed an inconvenient truth: The Protocol had little to do with saving the planet and everything to do with speeding development in Third World countries.

To remove this fatal flaw, during the UN's December 2009 Climate Conference in Copenhagen, Denmark, most Third World countries signed commitments to reduce greenhouse gas emissions. The problem most media outlets failed to mention was that these commitments were voluntary pledges. Unlike The Protocol itself, these commitments were not enforceable treaties. The flaw remained, but that didn't stop the US and other developed countries from pledging to fund a $100 billion per year aid package to help undeveloped countries install clean energy resources. The UN goal of spurring Third World development through partnerships with developed nations is working.

Competitive Advantages

Beyond the aid package mentioned above, The Protocol provided two competitive advantages for undeveloped nations. First, it increased the cost of energy in developed nations. Second, it provided incentives for developed nations to fund clean energy projects in Third World countries.

Energy is a commodity that underlies the overall cost of doing business, the cost of creating jobs, and the cost of practically everything we consume. By forcing developed nations to replace portions of their fossil fuel infrastructure, Kyoto increases the cost of energy in developed nations. For example, the average price of electricity in European countries that signed The Protocol is now two to three times higher than the US average.[252] Although there are many contributing factors, the cost of meeting mandated carbon reductions is a significant cost adder for Europe.

Meanwhile, The Protocol allows Third World nations to expand their use of low-cost, fossil-based energy. As a result, increased development and the transfer of both jobs and wealth should begin to flow to undeveloped nations.

The second competitive advantage provided by the Kyoto Protocol stems from certificates called Clean Development Mechanisms (CDMs). Each CDM certificate earned by a developed nation allows that nation to emit an extra ton of CO_2 back home. The Protocol allows developed nations to earn CDMs by investing in

---

252. Lindsay Wilson, "The Average Price of Electricity, Country by Country," September 25, 2013, http://theenergycollective.com/lindsay-wilson/279126/average-electricity-prices-around-world-kwh (Accessed 1-2-13).

clean energy projects in Third World Countries. The transfer of wealth continues.

Here's how CDMs work. When a developed nation invests in a clean energy project in the Third World, the UN assumes that each unit of clean energy produced will avoid CO_2 emissions that would have otherwise occurred. For each ton of avoided CO_2, the developed nation is granted a CDM allowing more CO_2 emissions back home. The net reduction in emissions is initially zero!

To their credit, the UN does limit the number of CDMs developed nations can use, so The Protocol eventually reduces overall CO_2 emissions. The urgency of those reductions, however, is obviously lacking.

The Third World country gets a shiny, new electric generator to power new job-creating industries, and the developed nation gets a high-priced sheet of paper that says it's okay to emit greenhouse gases in their own country. It's a wonderful mechanism for spurring development in Third World countries, but it doesn't exactly contribute to job creation in developed nations. It also does little to save the planet. Prior to the 2009 Climate Conference, European nations that signed The Protocol had already invested billions of Euros in such projects. The transfer of wealth is well underway.

The UN's solution to global warming is purely political, it ignores well over half of all human greenhouse gas emissions, and it ultimately targets less than 0.5% of the total greenhouse effect. The Protocol is specifically designed to speed development in Third World countries, not to save the planet. It's little more than a transfer of both wealth and jobs from developed nations to undeveloped nations.

The beauty of the Kyoto Protocol is in the marketing of the treaty. The marketing has convinced developed nations to send their money and jobs overseas in order to save the planet. Saving the planet is the last thing The Protocol was designed to accomplish.

Global warming is the crisis the UN won't let go to waste.

The Motives of China and Undeveloped Nations

China and India are considered undeveloped nations under the Kyoto Protocol—even though China was already the world's largest emitter of CO_2. China ratified the Kyoto Protocol in 2002, but that's done nothing to slow their greenhouse gas emissions.

Prior to the Great Recession, China was installing one new coal-fired power plant every week. They were installing the equivalent of the entire US coal-fleet once every three years.[253] Beijing alone was adding 14,000 new vehicles per day.[254]

Despite these facts, China has been insistent that every nation on the planet sign the Protocol. Here's what China had to say about the Protocol and the global warming debate:

> BEIJING (*Reuters*, 2009) - China called on rich nations to sign up to carbon emission cuts of 25–40 percent by 2020. . . . The [Chinese] official also said 'China, the world's top emitter of planet-warming greenhouse gases, wanted to commit to emissions reductions in certain industries but was still figuring out how to do this.'[255]

A similar story was carried by *Industrial Business Times*:

> China's top climate envoy . . . said it [China] would discuss a 2050 emissions goal [CO_2 reduction] only if rich nations offered more cash [to undeveloped nations] and carbon cuts. Xie Zhenhua said developed nations must commit to [CO_2]

253. Peter W. Huber, "We Cannot Make a Dent in Global Carbon Emissions," April 19, 2009, http://www.opposingviews.com/articles/opinion-we-cannot-make-a-dent-in-global-carbon-emissions (Accessed 12-29-10).
254. Jason Simpkins, "Saudi Arabia Agrees to Increase Oil Output After Crude Hits Another High," May 15, 2008, http://www.moneymorning.com/2008/05/19/saudi-arabia-agrees-to-increase-oil-output-after-crude-hits-another-new-high/ (Accessed 2-12-10).
255. David Stanway and Tom Miles; Editing by David Fogarty, "China calls for deeper CO2 cuts from developed world," (*Reuters*), May 12, 2009, http://www.reuters.com/article/2009/05/13/us-china-emission-cuts-idUSTRE54C0BE20090513 (Accessed 2-12-10).

cuts of at least 40 percent by 2020 from 1990 levels.[256]

It's easy to see why Third World nations have pushed for adoption of The Protocol. Although The Protocol has expired, these countries desperately wanted the US to ratify the treaty. They wanted it because then—and only then—would US taxpayers and US energy consumers begin sending billions of dollars overseas to obtain those high-priced sheets of paper called CDMs.

Third World countries have powerful motives for promoting the belief that humans cause global warming. We should also note that Third World nations comprise the majority of the UN's membership and are well represented in the IPCC. If someone is playing politics with climate change science, are we certain it's the Skeptics?

Global warming is the crisis Third World nations won't let go to waste.

The Motives of Climatologists

We're quick to disparage and mistrust corporate scientists because we know they have jobs to protect, profits to secure, and a boss to please. Our perception of independent scientists is much different. We see them as diligently pursuing the fundamental truth without bias and without the need to secure profits, protect their jobs, or keep the boss happy.

That may be our perception, but independent scientists have to earn a living like everyone else. If they don't work for corporations, they earn their living by teaching, consulting, and securing research grants. According to the Congressional Budget Office, annual US spending for climate studies rose from $4.0 billion in 1998 to $7.5 billion in 2009[257] and $8.8 billion in 2010.[258] The US alone spent

---

256. (*Reuters*) "EXCLUSIVE: China calls for more emissions cuts from U.S.," *Industrial Business Times* (December 9, 2009), http://www.ibtimes.com/exclusive-china-calls-more-emissions-cuts-us-352119, (Accessed 8-17-10).
257. US Congressional Budget Office (CBO), "Federal Climate Change Programs: Funding History and Policy Issues" (March 2010), http://www.cbo.gov/sites/default/files/cbofiles/ftpdocs/112xx/doc11224/03-26-climatechange.pdf (Accessed 3-15-12).

some $107 billion on climate research from 2003 to 2010. Someone is earning a lot of grant money studying climate change. Is global warming the dream crisis climatologists won't let go to waste?

Motives of the Environmental Lobby

Do we even need to comment on why the environmental lobby wants to promote belief in human-caused global warming? We can't harvest fossil fuels without disturbing the planet, and we can't burn fossil fuels without pollution and without emitting CO_2. While we can remove the majority of the worst pollutants from fossil fuels, there are no proven, commercially viable technologies to remove CO_2. The only way we can currently reduce CO_2 emissions is to reduce our use of fossil fuels—and there's not an environmentalist on the planet opposed to that result.

Widespread belief in human-caused global warming is the environmental lobby's utopia, their nirvana, and their Garden of Eden. It's their World Series, Super Bowl, NBA Championship, and World Cup, all wrapped up in one package. If they can convince the world that CO_2 is a pollutant, they will have accomplished more with one crisis than they've accomplished in their entire history of activism. Is that a good thing or a bad thing?

Either way, global warming is the crisis they can't let go to waste.

Motives of Al Gore

We've already mentioned that Al Gore has been called the world's first global-warming billionaire. Although that claim is an exaggeration, earning speaking fees in excess of $100,000 per lecture seems like a powerful motive. Growing one's net worth from some $2 million to $100 million is also more than ample motive. Global warming is the crisis that allowed Mr. Gore to increase his fame and fortune while gaining the admiration of millions around the globe. It's the crisis Mr. Gore didn't let go to waste.

---

258. Larry Bell, "The Alarming Cost Of Climate Change Hysteria"(8-23-11), *Forbes*, http://www.forbes.com/sites/larrybell/2011/08/23/the-alarming-cost-of-climate-change-hysteria/2/ (accessed 11-5-12).

Motives of Wall Street

Nearly every piece of climate legislation considered by the US Congress has included a cap-and-trade program for greenhouse gas emissions (CO_2). Although the program doesn't currently exist, we can envision Wall Street's interest by understanding how this program would work. In a cap-and-trade market, the government issues certificates called Allowances that operate like CDMs. Businesses and industries must submit one Allowance for each ton of CO_2 they emit. The government limits the number of available Allowances and establishes a free market exchange where Allowances can be bought and sold. Wall Street would operate this free market exchange.

The role of Wall Street was discussed in an article written by Matt Taibbi titled "The Great American Bubble Machine," which appeared in the July 13, 2009 edition of *Rolling Stone*. The article focused on the Wall Street profiteering that would accompany a climate cap-and-trade program. The article concluded that Wall Street's participation would only serve to increase the program's cost by adding the cost of sales commissions and allowing speculators to game the market.

David Sokol, Chairman of MidAmerican Energy Holdings Company, explained the impact of Wall Street profiteering in an editorial called "Let's Have Cap and No Trade," which appeared in the May 19, 2009 edition of the *Washington Post*:

> If you liked what credit default swaps did to our economy, you're going to love cap-and-trade. Just read Title VIII of the bill [the American Clean Energy and Security Act of 2009], which lets investment banks, hedge funds, and other speculators participate in the [climate] cap-and-trade market. They don't have emissions to cut; they have commissions to make.[259]

---

259. David Sokol, "Let's Have Cap and No Trade," *Washington Post* (May 19, 2009), http://articles.washingtonpost.com/2009-05-19/opinions/36804915_1_carbon-dioxide-greenhouse-gas-emissions-cap-and-trade-market (Accessed 3-11-10).

Even more disturbing, as described by Louis Redshaw, head of Environmental Markets at Barclays Capital, the proposed climate cap-and-trade market would become the largest commodity market in the world.[260]

In other words, CDM-like certificates for CO_2 are poised to become a bigger market than gold, oil, and yes, even pork bellies. Keep in mind that a ton of emitted CO_2 is not a commodity like wheat or gold that actually serve a purpose or provide value. Because nothing of value will be traded, who will provide the funds to build this new, world's biggest commodity market? The answer is obvious—every consumer living under a carbon cap-and-trade mandate. Each time we buy gasoline, pay our heating and cooling bill, fire up the lawnmower or the barbeque grill, turn on the television, or buy anything that requires energy to plant, grow, harvest, manufacture, transport, or store; we'll be paying to help create the world's largest commodity market.

Wall Street traders, investment bankers, commodity brokers, and financial speculators all have a vested interest in promoting the belief that human activities cause global warming. It's the crisis they won't let go to waste.

Government Motives

For governmental leaders, climate legislation provides a rare opportunity: The chance to pass a tax increase that voters actually support. It won't be called a tax, of course. Instead, it'll be called "a pollution reduction fund" or "cap and cash back"—just as the marketing consultants recommended. Regardless of what it's called, climate legislation equates to a hidden tax on practically everything we buy and use.

President Obama's original budget for 2012-2019 included $646 billion in new revenue from a proposed CO_2 cap-and-trade

---

260. James Kanter, "Carbon trading: Where greed is green," *New York Times* (June 20, 2007),
http://www.nytimes.com/2007/06/20/business/worldbusiness/20iht-money.4.6234700.html?pagewanted=all&_r=0 (Accessed 3-11-10).

program.[261] That's some $80 billion per year that would have come from your pocket and mine if climate legislation had been signed into law. It's a new source of revenue many in Congress would love to control.

Beyond this obvious motivator, congressional leaders understand the voting power of the green lobby. Alienating this lobby spells doom at election time. Congress also understands how the media would portray their actions if they dared to oppose the findings of the "overwhelming majority of scientists." It would be far worse than the ridicule Columbus faced when he claimed the world wasn't flat—and he was almost imprisoned for spreading such blasphemy!

Members of Congress have ample motive for supporting belief in human-caused global warming. It's the crisis they can't let go to waste.

Media Motives

I'm frankly at a loss to explain why the popular media has continued to present a one-sided, biased view of global warming. Many point to the media's liberal bias, which is based on the stereotypical view of journalists as longhaired, hippie-types who are among the greenest of the green. That could be one reason, but there's another more traditional motive: profit.

The news media is not in business to deliver facts. They're in business to sell advertising. The more they increase subscription rates or increase viewers and listeners, the more they can charge for airing commercials. The news itself may be a public service, but the bottom line for the news media is the same as it is for any major corporation: making a profit. The news is designed to shock, surprise, and entertain—all in an attempt to increase audience size and increase profits. That's why catastrophe, scandal, corruption, and environmental wrongdoing get top billing. Mundane facts don't sell newspapers. The facts need not be spoken if it means selling

---

261. (*Reuters*) Jeff Mason and Timothy Gardner, "Obama budget drops revenue outlook for carbon trade" (February 1, 2010), http://www.reuters.com/article/2010/02/01/us-obama-budget-climate-idUSTRE6101VB20100201 (Accessed 9-7-10).

more advertising.

Whatever the media's motivation—whether it's promoting all things green or sensationalizing the news for profit—the media has failed to deliver an unbiased accounting of global warming.

Media Bias

During the UN's 2009 Climate Conference, newspapers around the globe carried daily front-page articles detailing the dire consequences if we fail to curb emissions of CO_2. Televised evening news programs also featured stories about global warming. One in particular caught my attention. It aired on December 7, 2009, the opening day of the Climate Conference. This news segment focused on Bangladesh.

If you know anything about Bangladesh, you already know this story was biased from the beginning. Most of Bangladesh is less than ten meters above sea level. In this case, the reporter made his broadcast from the southern coast, which is the lowest part of the country, where monsoons, floods, and cyclones regularly inflict heavy damage. In fact, roughly 40% of the world's tropical storm surges occur in Bangladesh.[262]

At the beginning of the story, the commentator told the audience everything they needed to know about the rest of the story. He casually mentioned that the sea walls protecting this area had recently been destroyed by a cyclone. The rest of the coverage included images of flooded shacks and starving children standing in knee-deep water. The report made it clear: This will be the world's fate if we don't stop using fossil fuels.

This report was so misleading it wasn't news at all. This was the story of a land that flooded because their protective sea walls collapsed, not because we burn fossil fuels.

---

262. Ubydul Haque, Masahiro Hashizume, Korine N Kolivras, Hans J Overgaard, Bivash Das, & Taro Yamamoto, "Reduced death rates from cyclones in Bangladesh: what more needs to be done?" (February 2011), http://www.who.int/bulletin/volumes/90/2/11-088302/en/ (Accessed 10-23-13).

Conspiracy Anyone?

We generally scoff at claims made by conspiracy junkies. Our reaction to the mere suggestion of a global warming conspiracy might sound something like this:

> Congratulations! You've blown the lid off a conspiracy coordinated around the globe by thousands of independent scientists, numerous heads of state, journalists, and public interest groups, not to mention a billion or so normal citizens who understand the dangers of greenhouse gases. We can all sleep better at night knowing you've exposed this diabolical conspiracy.

Yes, the required scope of such a worldwide conspiracy makes the notion laughable, but at the same time, how else do we explain the volume of facts that have gone unreported? As improbable as it may sound, it's clear that someone has successfully hidden critical facts and presented a totally one-sided view of global warming. Although we find it difficult to imagine, the results certainly have all the ingredients of a conspiracy.

Many of us play an unwitting part in this improbable conspiracy. We're convinced that we face a genuine, human-caused catastrophe. As a result, we're doing whatever we can to help. Along with the media, environmental organizations, and others, we're helping spread the word to promote smarter energy use. Who could blame us? For one thing, it's the right thing to do whether or not it has an impact on climate. Beyond that, everything we've been allowed to consider regarding Earth's climate tells us that we're responsible for an unfolding climate nightmare. We're trying to do the right things, as we should, but the fact remains, we haven't been given all the facts and many of our proposed actions will cause more harm than good.

Regardless of what we call it, somewhere in the upper echelons of power inside the UN, the environmental lobby, the media, and world governments, someone is screening the information we're allowed to see. Someone is creating Nobel Prize-winning documentary films filled with false claims and bad science. Others are promoting news stories that ignore critical facts. If that doesn't equate to a conspiracy on some level, I'm not sure what does.

Means and Opportunity

We've established numerous motives, so let's move to means and opportunity. If you've seen the movie or read the book *Wag the Dog*, you've seen a perfect example of how entire societies can be fooled. The title suggests that the dog isn't in control of the tail. Instead, the tail is controlling the dog. With global warming, society is the dog, and the UN, the environmental lobby, and the media are the tail, controlling our beliefs.

The *Wag the Dog* storyline goes like this: Faced with scandal and the loss of popular support, America's President hires a marketing guru to increase voter support. The marketing guru suggests staging a fake war. Together, the guru and the President choose a man to serve as the war's hero, they create a courageous storyline for the hero, and they film the entire war in the basement of the White House.

The media, in their zeal to sensationalize a hot story, buy into the fake war and the bogus hero. News of the war dominates the headlines as the nation unites in a common cause. Voters quickly forget the President's shortcomings and almost everyone supports the war because it's the right thing to do.

What may have seemed far-fetched fiction in 1997 has become today's reality. Today we have a make-believe war to unite us against a common enemy—CO_2. We have a handpicked made-for-television-and-the-movies hero—Al Gore. We also have secondary heroes, such as the UN, the IPCC, and the environmental lobby—all striving to reveal the inconvenient truth. Finally, we have a media eager to sensationalize the story.

Today's *Wag the Dog* war isn't a war of bullets. It's a war of influence being conducted by marketing gurus, it's a war designed to shape our beliefs, and it's been a rousing success.

Dress Rehearsals

Pulling off a *Wag the Dog* conspiracy in the movies is one thing, but it can't be that easy to accomplish in real life, can it? The reality is environmentalists, the UN, and world governments have had a number of dress rehearsals for accomplishing this unlikely feat. The worldwide ban on DDT was the first rehearsal.

Today we all understand the need to curb acid rain, we know about the hole in the ozone, and we all want to save the whales. We understand these things because they've been presented to us in precisely the same way global warming has been presented.

The processes involved, the steps required, and the groups involved are all the same for each of these issues: 1) Identify a potential threat—whether real or imagined, 2) Develop data supporting this potential threat, 3) Hire marketing gurus to put the proper spin on the threat, 4) Use the media and the vast network of environmental organizations to publicize the threat, and 5) Keep repeating the threat so often it becomes accepted as the truth.

This isn't to say that acid rain or the hole in the ozone weren't legitimate issues. It simply shows that the processes to create global deceptions have been proven to work time and again by the same groups promoting human-caused global warming. A worldwide conspiracy to fool the masses is much easier to accomplish than we'd like to believe.

Perhaps the best example of a dress rehearsal for global warming involved the hole in the ozone. The ozone in question is the protective layer of ozone in the upper atmosphere that blocks harmful cosmic rays (UV-B) from reaching Earth's surface. In September 1987, the UN organized a world conference in Montreal, Canada, to address threats to this protective layer of ozone. The conference resulted in the Montreal Protocol: an international treaty designed to reduce emissions of Chlorofluorocarbons (CFCs). Sound familiar?

CFCs are chemical compounds used mainly in refrigeration and air conditioning, although they were also in widespread use as accelerants in aerosols such as hair spray and spray paint. CFCs are safe, stable, nontoxic, non-corrosive, and nonflammable chemicals, which made these compounds uniquely useful in many consumer and industrial applications.

Concern over the role of CFCs actually began in 1973 with studies conducted by two scientists from the University of Michigan, Richard Stolarski and Ralph Cicerone. These researchers theorized that chlorine from rocket fuel could unleash a complex chain reaction in the stratosphere that would destroy ozone over a period of decades. Since very little free chlorine existed at the required altitude and emissions from rockets were small, the study raised little concern and was largely ignored.

Concern grew roughly a year later when two researchers from the University of California, Irvine, Mario Molina and Sherwood Rowland, discovered that CFCs could migrate slowly up to the stratosphere, where radiation could break chemical bonds and release large volumes of free chlorine.[263] Based on the results of the earlier University of Michigan study, it was theorized that this free chlorine could destroy ozone in the stratosphere. This theory was also largely ignored because there was no evidence of ozone depletion, no evidence that the theory worked in nature, and ozone levels actually increased in the 1960s despite the rapidly expanding use of CFCs.

According to author Brien Sparling, "A large shock was needed to motivate the world to get serious about phasing out CFCs. That shock came in a 1985 field study by Farman, Gardinar, and Shanklin."[264] The referenced field study indicated that ozone levels over the Antarctic were well below normal January levels. This finding became popularly known as the hole in the ozone. When combined with the two previous studies, the stage was set for international action to stop the use of CFCs.

According to Richard E. Benedict, the lead US representative involved in negotiations for the Montreal Protocol, many uncertainties remained in 1986 when negotiators began their discussions in Montreal.[265] For example, no significant ozone depletion had been detected in over thirty years of measurements. This finding was in stark contrast to computer models that were predicting severe depletion. Scientists realized that their computer

---

263. Richard E. Benedict, "Science, Diplomacy, and the Montreal Protocol," from *The Encyclopedia of Earth* (June 12, 2007), http://www.energy.probeinternational.org/climate-change/the-deniers-and-promoters/science-diplomacy-and-the-montreal-protocol-0 (Accessed 12-13-09).
264. NASA, Brien Sparling, "Ozone Depletion: History and Politics," http://www.nas.nasa.gov/About/Education/Ozone/history.html (Accessed 4-28-10).
265. Richard E. Benedict, et al.

models were wrong and many concluded that the threat from CFCs wasn't supported by the facts. Today, we know that changing solar winds and cosmic rays may be the real cause of the hole in the ozone. Such mundane issues, however, didn't stop the media from sensationalizing the story. These issues also didn't stop the public's fear, and they didn't stop the Montreal Protocol from being implemented.

Many would argue that it doesn't matter whether our scientists were right or wrong about CFCs because waiting for proof may have delayed action until it was too late to prevent disaster. Many would similarly argue that global warming must be approached with the same level of caution. While there are strong proponents for immediate action on climate change, we have to realize that chorine and CO_2 are vastly different operators. To the best of our knowledge, high levels of chlorine in the stratosphere don't occur naturally. Therefore, if we're putting chlorine in the stratosphere, it could pose problems. Conversely, CO_2 is naturally occurring; it's essential to life on this planet; and it has been in our atmosphere at far higher levels than anything contemplated to occur due to our use of fossil fuels (and ice ages still occurred). History also shows that CO_2 is a climate follower, not a climate driver. There are ample reasons for treating greenhouse gases and CFCs differently.

This discussion isn't meant to criticize our immediate action on CFCs. The story is important because the ban on CFCs was the perfect dress rehearsal for global warming. Convincing societies around the world that humans cause climate change is far easier to achieve than we'd like to believe. The required steps have been tested and proven time and again in recent history. All you need is the right combination of trusted organizations; *independent* research; supportive computer models (even if they're wrong); a profit seeking, sensationalizing media; and, of course, motive.

As absurd as conspiracy may sound, perhaps it's time we realized that our belief in human-caused global warming is a perfect fit for the term.

Chapter 19
Vapor Proof

"Only an insignificant fraction of scientists deny the global warming crisis. The time for debate is over. The science is settled."
—Al Gore (1992)[266]

As previously mentioned, it's been said that an organized few can control the disorganized many. That's especially true if the organized few have the media on their side.

The media has ignored essentially every challenge raised by independent (disorganized) Skeptics. At the same time, the media has embraced the IPCC and essentially every comment or quote supporting human-caused global warming. What the media hasn't done, however, is provide an unbiased, in-depth review of the science behind the IPCC's *proof*.

In this chapter we'll explore key assumptions used by the IPCC and the processes they followed to reach their conclusions. Because the IPCC's proof ultimately relies on computer models, let's begin by exploring the fundamental rules of computer modeling.

The Rules of Computer Simulation Modeling

A great deal of my career was spent working side by side with computer simulation modelers. I actually ran computer simulation models for a few simplistic evaluations and quickly decided that I didn't want to become a computer modeler. The work was far too repetitious for my taste. Instead, I played a role similar to the role played by geologists. I provided the computer modelers with the data they needed to help their models predict the future. While

---

266. "They call this a consensus?"(June 19, 2007), *National Post*, http://www.financialpost.com/story.html?id=c47c1209-233b-412c-b6d1-5c755457a8af (Accessed 10-23-13).

working side by side with computer modeling experts, I learned the first, second, and third rules of computer simulation modeling.

The first rule tells us everything we need to know. The first rule of computer simulation modeling is simply this: "All models are wrong!"

Obviously, this is a tongue-in-cheek rule seldom shared outside the close-knit community of computer modeling experts. Computer models are tremendously powerful tools, but their results are nothing more than probabilistic, statistical analyses that are totally dependent on the assumptions used in the model. If computer models can't accurately predict the next flip of a coin, there's little hope that more complex models will provide greater accuracy. The first rule of computer modeling should come as no surprise.

Garbage In

We all know the second rule of computer modeling, which is "Garbage in gives you garbage out." You don't need to be a rocket scientist to know that a model's results are only as good as the formulas, assumptions, and probabilities that are used in the model. Due to the vast number of uncertainties involved and because clairvoyance isn't one of the prerequisites for becoming a climatologist, we shouldn't expect climate models to be particularly accurate.

Finding the Right Rock

The third rule of computer modeling is called "Finding the Right Rock." This is the one rule of modeling the IPCC really doesn't want us to understand. Computer modelers describe this rule by saying, "If you tell me which rock you want me to find, I can find it."

The term *rock* is a play on words. It's as though the modelers were hired to search through a huge pile of rocks (data) in order to find the right rock (the answer the boss wanted them to find). After years of applying their trade, expert modelers learn which assumptions to change to produce the desired result. They know how to find the right rock.

Finding the right rock is easy, but the peer-review process is supposed to limit the games that can be played. Before a model's results are accepted, *independent* subject matter experts review and either approve or reject the assumptions used in the model. While this review is supposed to limit the modelers' ability to bias their findings, in my experience, it never stopped the modelers from finding the right rock.

As previously discussed, the purity of the IPCC's peer-review process is strongly challenged by many of the IPCC's own scientists. More importantly, the IPCC peer-review was ultimately conducted by a small, select group of government representatives who edited the final results. If this select group didn't want to let a good crisis go to waste, their editing could easily ensure that the IPCC found the right rock.

Climategate

It's difficult to imagine climate scientists deliberately tweaking their assumptions to find the right rock (obtain the desired answer). However, Climategate allegedly demonstrated this exact behavior. For those not familiar with the story, Climategate exposed emails from noted IPCC scientists that were intercepted (*hacked* was the preferred term for many media outlets) and made public. The emails reflected a biased selection of data by IPCC scientists and the exclusion of data that didn't support the UN's desired story. There were at least two separate Climategates: Climategate 1.0 (2009) and Climategate 2.0 (2011). [267, 268, 269]

If you were already a Skeptic, Climategate proved that facts were being manipulated to tell the desired story. If you were already a Believer, Climategate proved nothing. In fact, many environmental

267. John R. Lott, "The Next Climategate" (February 10, 2010), *FoxNews.com*, http://www.foxnews.com/opinion/2010/02/09/john-lott-joseph-daleo-climate-change-noaa-james-hansen/ (Accessed 12-20-11).
268. James Heiser, "NOAA and the New Climategate Scandal," *The New American*, http://www.thenewamerican.com/tech-mainmenu-30/environment/2930-noaa-and-the-new-qclimategateq-scandal (Accessed 12-20-11).
269. Juliet Eilperin, "'Climategate' resurfaces with a new round of e-mails," *Washington Post* (November 23, 2011), http://www.washingtonpost.com/national/health-science/climate-gate-resurfaces-with-a-new-round-of-e-mails/2011/11/22/gIQAGBcAmN_story.html (Accessed 12-21-11).

websites and several media stories referred to Climategate as a conspiracy concocted by Skeptics in an attempt to divert attention away from the real science and the overwhelming body of evidence linking humans to global warming.[270]

A simple web search for "Climategate emails," will show a bounty of discussions both supporting and refuting the significance of the leaked emails. For example, *Factcheck.org* indicates the emails were taken out of context and don't show any real deception. Other sites claim to have investigated the emails and found no wrongdoing. We generally trust fact-checking websites like *Factcheck.org* and *snopes.com*, but if you also search the web for "fact check bias," you'll find that most of these *trusted* websites are biased and dedicated to promoting either a liberal or a conservative slant on the issues they investigate. Once again, you and I are left trying to decide what we can and can't believe.

As always, the best way to know what we can or can't believe is to learn the fundamental facts. In this case, reading a few of the emails is the best way to decide what to believe. You can read portions of the hacked emails at the website footnoted at the end of the list below. Admittedly, many of the *hacked* emails are difficult to translate and not all of the 220,000 emails demonstrate intentional deception. There are, however, emails that are clear. A few excerpts from those emails are paraphrased below:

- I'm personally very skeptical of our climate recreations, yet I have to sound like a highly qualified supporter of human-caused global warming.
- The decision to lower the sun's climate impact is primarily based on one paper, despite numerous studies reaching the opposite conclusion.
- The political interests are huge. We have to deliver a strong story supporting human-caused global warming because governments don't want to look foolish.

---

270. John Cook, "The question that skeptics don't want to ask about 'Climategate'" (November 18, 2010), *Skeptical Science*, http://www.skepticalscience.com/The-question-that-skeptics-dont-want-to-ask-about-Climategate.html (Accessed 9-9-11).

- This paper is too optimistic [it doesn't emphasize the role of CO_2]. We'll have to cut some of the conclusions [to show CO_2 as the primary cause].
- Observations do not show rising temperatures unless you accept one single study and discount a wealth of others.
- I agree the science is manipulated to put a political spin on the story.
- The big decisions are being made at the last hour by a small core group.
- These results are deceptive. A large number of misleading presentations are being used by the IPCC.
- We have to convince readers there has been an increase in our understanding, but I can't tell you what, if anything, has increased our understanding.
- We have to make sure they're losing the PR battle.
- A great deal of today's warming is explained by changes in the sun alone. [271]

It would seem these emails are more than newsworthy, but instead of becoming the scandal of the century—*IPCC Scientists Caught in Conspiracy*—Climategate quickly faded from public view. The media gave the UN and their scientists a free pass. Climategate 2.0 was barely mentioned in the popular media. Apparently, all has been forgiven.

Never mind the fact that IPCC scientists were caught cooking the books. Forget the open letters signed by skeptical scientists from around the world. Ignore the Senate Minority Report, the UN's singular goal of spurring development of undeveloped nations, and Earth's history that refutes most of the claims we've heard regarding global warming. Forget that temperature has always changed before CO_2 levels changed and CO_2 levels many times higher than today couldn't stop ice ages from occurring. Forget all that: It's those

---

271. Archive-org.com, "Climategate 2.0 FOIA 2011 Searchable Database," http://archive-org.com/page/3084861/2013-10-26/http://foia2011.org/ (Accessed 7-16-13).

darned Skeptics we can't trust!

It's truly an incredible story. Some of the IPCC's top scientists were caught clearly stating they were having difficulty supporting the desired story—and the media portrays it as a conspiracy concocted by the Skeptics!

How IPCC Scientists Found the Right Rock

We've just seen a Climatgate email indicating that the decision to lower the sun's impact on climate was based on one cherry-picked report at the exclusion of numerous studies reaching the opposite conclusion. This decision was one of the most critical assumptions used by the IPCC to find the right rock. It wasn't, however, the most important assumption made by the IPCC.

The most important assumption indicated that human emissions of CO_2 create a feedback that increases the direct impact of CO_2. This feedback assumes that CO_2 causes temperatures to increase, which then increases evaporation, which then increases atmospheric water vapor, which then causes further increases in temperature. The IPCC also assumed this process is self-sustaining, which resulted in a multiplying effect to greatly increase the actual impact of CO_2. This feedback makes sense because we know that warm air retains more moisture than cold air. That's why relative humidity is higher in summer and lower in winter.

The IPCC also assumed that this feedback worked in only one direction: it feeds upon itself to greatly increase temperature. Perhaps this explains why the IPCC models have significantly overstated warming for the last twenty years. Many scientists object to the assumption of a one-way feedback. These scientists say that negative feedbacks are also created. A negative feedback would act to cool the atmosphere. For example, increased evaporation will lead to increased cloud formation and increased rainfall (we've already seen this fact being used by Believers to explain why expanding glaciers can indicate global warming). Increasing clouds will reflect more of the sun's energy back into space and rain has a natural cooling effect on the atmosphere. According to NOAA,

> As water vapor increases in the atmosphere, more of it will eventually also condense into clouds, which are more able to

reflect incoming solar radiation [thus allowing less energy to reach the Earth's surface].[272]

Dr. Roy Spencer agrees with NOAA. Dr. Spencer is a Ph.D. in meteorology, a Principal Research Scientist at the University of Alabama in Huntsville, and a Senior Scientist for Climate Studies at NASA's Marshall Space Flight Center.

In his book, *Climate Confusion—How Global Warming Hysteria Leads to Bad Science, Pandering Politicians, and Misguided Policies that Hurt the Poor* (page 71), Dr. Spencer cites numerous concerns regarding the IPCC's feedback assumption. One of his arguments indicates that the level of solar energy reaching Earth is enough to saturate our atmosphere with water vapor in roughly one week's time. That doesn't happen because natural processes—such as cloud formation and rain—act to maintain atmospheric relative humidity far below the saturation point. Since cloud formation and rain act to cool the atmosphere, Dr. Spencer and many others conclude that an increase in temperature—whatever the cause—will produce negative feedbacks that counteract warming.

Data from weather balloons and NASA satellites agree with Dr. Spencer. A report from NOAA indicates that water vapor in the stratosphere *declined* 10% between 2000 and 2010.[273] This may explain why atmospheric cooling has occurred since 1998. It may also explain why IPCC computer models have significantly overestimated warming for the last twenty years.

More importantly, this singular finding disproves the IPCC's most critical modeling assumption regarding CO_2's role in climate change!

---

272. NOAA, National Climate Data Center, "Greenhouse gases, Water Vapor," http://www.ncdc.noaa.gov/cmb-faq/greenhouse-gases.php (Accessed December 2, 2009).

273. Susan Solomon, et al., "Stratospheric Water Vapor is a Global Warming Wild Card," NOAA (January 28, 2010), http://www.noaanews.noaa.gov/stories2010/20100128_watervapor.html (Accessed 10-19-13).

The Bottom Line

The *undeniable proof* that humans cause global warming ultimately rests on a computer model and two critical assumptions: The sun is a minor player in climate change, and CO_2 causes a one-way feedback. Both of these assumptions violate real-world observations as well as numerous scientific studies, but again, we won't hear this from the mainstream media.

Who are the true independent scientists in this debate? Is it Bjorn Lomborg and the other Skeptics, or is it the IPCC's government appointed representatives reaping the rewards of hundreds of billions in research funding? Can we trust the IPCC's proof? Can we trust the last twenty-plus years of sensational media headlines? Is the science complete? Have we been told the whole story?

You'll have your own answer to these questions, but two things should be crystal clear. First, the motives of those promoting human-caused global warming are phenomenally powerful. Second, we haven't been told the whole truth and nothing but the truth—and that's a far greater concern than choosing sides in the global warming debate.

We should never underestimate the power and influence of the environmental lobby. The motives of the UN should not be forgotten. The media's ability to influence our beliefs should terrify us all.

Believe what you want regarding the cause of today's warming, but believe this: many of our most strongly held beliefs regarding climate change (and a host of other critical issues) are little more than popular deceptions.

Part IV
We're from the Government,
We're here to Help

"Any man who thinks he can be happy and prosperous by letting the
Government take care of him; better take a closer look at the
American Indian."
—Henry Ford

Chapter 20
Energy, the Final Frontier

"Energy cannot be created or destroyed."
—The First Law of Thermodynamics

Humans require food, water, and shelter for survival. Energy is likely the next priority on our list of basic survival needs. Running out of energy is not an option, yet we know we'll eventually deplete our finite supply of fossil fuel. Consequently, one of the most critical hurdles facing humanity is developing sources of energy that don't rely on limited resources.

The questions surrounding this hurdle are significant. How much time do we have before we run out of fossil fuel? Are today's renewable technologies the answer we need? Is our current approach—limiting CO_2 emissions and subsidizing renewable energy—a wise approach? Are we investing in the right technologies? Do we understand the pros and cons of our choices?

The Best Clean Energy Investments

The best clean energy alternatives aren't the sexiest new technologies we can imagine. They're actually the simplest alternatives we know.

Insulation. This isn't sexy, but it's the best and the most economical clean energy investment we can make. It eliminates our most inefficient use of energy—wasted energy—and it eliminates that need forever! Insulating our homes, offices, and factories to recommended levels is the first step we should take in our quest to use energy wisely. Obviously, there's a point of diminishing returns: a point where adding more insulation is no longer economical, efficient, or environmentally friendly.

Passive Solar. Like insulation, this technology is about as simple as it gets. If you live in the Northern Hemisphere and build your home with south facing windows, the winter sun passing low on the southern horizon will provide warmth to offset your heating bills. Proper window tinting and roof overhangs will shade windows from the summer sun to keep air conditioning bills low. If windows facing south aren't an option, you can install Trombe walls. These are dark, hollow chambers with an exit vent at top (going into the home) and an inlet vent at the bottom (taking air out of the home). When the sun hits the dark outer surface of a Trombe wall, the air inside is heated and natural convection circulates warm air into the home. You won't win any curb appeal contests, but you'll heat much of your home with free energy. Just be sure to close the vents at night and in summer.

Efficiency. Investing in more efficient motors, lights, cars, and appliances is better than investing in new sources of energy production. It's even better than investing in renewable energy. Why? Like proper insulation, improved efficiency eliminates the need for wasted energy, and it eliminates that need forever. It's better to eliminate the need for energy when and where it's feasible and economic, rather than investing in a new source of energy production.

We should note that increased efficiency is already paying dividends. For oil usage, our vehicles are getting more miles per gallon and it's making a difference in our demand for oil. For electricity and natural gas, efficiency improvements are also beginning to show results. Air conditioning efficiency in particular is having a dramatic impact on peak electric usage. In the 1960s, air conditioning was a novelty for many homes. Between 1960 and roughly the year 2000, much of our increased use of electricity was driven by the addition of air conditioning. Today, most air conditioning installations are replacements of old, inefficient units with systems that are significantly more efficient. Electric utilities are seeing the impact through slower load growth, and in some cases, annual electric usage is actually declining—although part of the decline is also attributed to a slow economy.

The arrival of LED lights will also greatly reduce energy usage. For example, my hometown was the first city with a population over 100,000 to change all of its streetlights to LED's. This project will save over 5 million kilowatt-hours of electricity per year!

Conservation. If you have teenagers at home, you know they're among the greenest people on the planet. If your experience is like mine, you also know that teenagers understand the "on" button, but haven't figured out the "off" button. When I get home from work, it seems every light in the house is on, at least one TV is blaring, and the stereo is blasting-away upstairs in our teenager's room. Meanwhile, our teenager is down the street visiting friends. Teenagers may be green, but their carbon footprints are spectacular—and that's before they get their driver's license. The point is energy conservation is too simple to be ignored. Like any pursuit, however, conservation can be taken to extremes that hurt the economy, erode standards of living, and ultimately harm health (think of the elderly foregoing air conditioning on dangerously hot days). There is a point of diminished returns.

Fossil Fuels: How Much Time?

As recently as 2009, some projections indicated the world would run out of oil in less than 50 years, natural gas in less than 75 years, and coal in just under 120 years.[274] Such dire warnings have proven far too pessimistic. For example, in 1977, the US Energy Information Administration (EIA) indicated the US had only 32 billion barrels of recoverable oil and 207 trillion cubic feet of recoverable natural gas (ignoring coal reserves for the moment). Between 1977 and 2010, the US extracted nearly three times those totals, yet in 2010, reserves were 33% above the 1977 projections—and reserves have skyrocketed since then![275] The US now has more fossil fuel reserves than any nation on the planet.[276]

In fact, worldwide reserves are increasing, despite rapid growth in consumption. Discoveries in 2013 alone help explain why fossil

---

274. *Eco-Info.net*, "When Will We Run Out Of Fossil Fuels?" http://www.eco-info.net/fossil-fuel-depletion.html (Accessed 12-27-13).
275. Jeffrey Rissman, Ibid.
276. Gene Whitney, Carl E. Behrens, Carol Glover, "US Fossil Fuel Resources: Terminology, Reporting, and Summary" (November 30, 2010), the Congressional Research Service (CRS), http://epw.senate.gov/public/index.cfm?FuseAction=Files.view&FileStore_i d=04212e22-c1b3-41f2-b0ba-0da5eaead952 (Accessed 2-9-13).

fuel reserves are expanding rather than shrinking. In 2013, Australia announced a newly discovered oil and gas reserve that rivals Saudi Arabia,[277] major new oil deposits were discovered in the Gulf of Mexico,[278] the Permian Basin in Texas was projected to be the second largest oil discovery in history,[279] and Brazil discovered new oil and gas deposits that may be the largest find of 2013.[280]

Despite these enormous discoveries, coal remains the world's most abundant fossil fuel resource, and the US holds over one quarter of the world's coal. The last fossil resource we'll exhaust is coal, yet it's the first one targeted by the environmental lobby. Environmentalists say we need sustainable energy. That's a goal we can all embrace, but the first step environmentalists propose toward reaching that goal is to stop using coal. How does walking away from the most abundant fossil fuel resource on the planet help sustain our ability to use energy? The environmental game plan makes no sense. It only serves to accelerate the day we run out of oil and natural gas. What's sustainable about following this course of action?

Beyond coal, oil, and natural gas, scientists are developing methods to recover frozen methane hydrates from Artic permafrost and deep oceans. This new fossil fuel resource is a frozen form of natural gas that holds more energy than all previously discovered

---

277. *CBSNews*, "Australian shale oil discovery could be larger than Canada's oilsands," January 24, 2013,
http://www.cbc.ca/news/business/australian-shale-oil-discovery-could-be-larger-than-canada-s-oilsands-1.1320034 (Accessed 12-14-13).
278. *Oil & Gas Journal*, OGJ Editors, "BP, ConocoPhillips make oil discovery in deepwater gulf," December 18, 2013,
http://www.ogj.com/articles/2013/12/bp-conocophillips-make-oil-discovery-in-deepwater-gulf.html (Accessed 12-18-13).
279. Wyatt Investment Research, "The 2nd Largest Oil Discovery in the History of the World," *NASDAQ*, October 02, 2013
http://www.nasdaq.com/article/the-2nd-largest-oil-discovery-in-the-history-of-the-world-cm282232 (Accessed 12-17-13).
280. Jeb Blount, *Reuters*, "New Brazilian oil discovery may be biggest find of the year." September 27, 2013, FinancialPost.com,
http://business.financialpost.com/2013/09/27/new-brazilian-oil-discovery-may-biggest-find-of-the-year/?__lsa=b658-beef (Accessed 12-14-13).

fossil fuel reserves combined![281]

The fact is we have a couple of centuries—and most likely far longer—before we'll face a true energy crisis. That's not an excuse to be cavalier with our use of energy and it doesn't mean we should abandon our quest for sustainable energy. It simply means we have time to plan for the inevitable and time to develop wise and meaningful energy policies. Are we taking the time to do so, or are we rushing to judgment based on sensational headlines and inconvenient truths?

The Unfair Advantage

Renewable energy is a wonderful thing, but like it or not, fossil fuels have an unfair advantage. We've spent untold trillions of dollars and well over 100 years developing the technologies and building the infrastructures that make fossil fuels cheaper, more convenient, and functionally superior to the sexiest new energy technologies existing today. Replacing that infrastructure won't be cheap, and no matter how hard we try, it can't happen overnight.

Walking away from fossil fuels will likely be the most costly endeavor humans have ever attempted. The faster we move, the more economically painful it will be. The high cost also places an emphasis on doing it right the first time. We don't want to replace our energy infrastructure twice.

We're told that this unfair advantage has disappeared. We're told we can no longer afford to use *costly* fossil fuels. Do the fundamental facts support this allegation?

Why We Burn Coal

As stated above, coal is the most abundant fossil fuel resource on the planet. It also provides the lowest cost electricity available today. That's the opposite of what we've been told, but if we ranked the average US retail price of electricity by state, the states at the bottom of our list—the states with the lowest average price for electricity—would be states where coal provides the majority of the

---

281. Geology.com, "Methane Hydrate The world's largest natural gas resource is trapped beneath permafrost and ocean sediments," http://geology.com/articles/methane-hydrates/ (Accessed 12-27-13).

electricity.[282] Knowing that, it's difficult to understand why we're being told we can no longer afford to rely on *costly* coal.

Coal represents 93% of total US fossil energy reserves.[283] What do you suppose will happen to the cost of electricity, oil, and natural gas once the EPA ensures we can no longer use 93% of our vast fossil reserves?

The Cost

It's been said that energy from the wind and sun are free. That's obviously not true or we'd all have wind turbines in our backyards and solar panels on our rooftops. The US Department of Energy (DOE) tells us that new, large-scale wind farms produce energy for around nine cents per kilowatt-hour ($0.09/kWh) and new, large-scale solar installations produce energy at a cost of more than fourteen cents per kWh ($.144/kWh).[284] The cost of small-scale (residential) wind and solar is much higher, which is the primary reason we don't all have wind turbines in our backyards and solar panels on our rooftops. By comparison, most existing coal-fired power plants produce energy for around three cents per kWh ($0.03/kWh). Even the oldest, most inefficient coal plants operating today produce energy for less than seven cents per kWh ($0.07/kWh).

These are not the cost comparisons we're accustomed to seeing. As advised by their marketing gurus, environmentalists avoid discussing the specifics and the costs because the facts "need not be spoken." There are, however, two costs that environmentalists will discuss: the *social cost* of fossil-based energy

282. US EIA, "Average Retail Price of Electricity to Ultimate Customers, by State," Electricity Data Table 5.6.B.
http://www.eia.gov/electricity/data.cfm#sales (Accessed 7-21-13).
283. Daniel O'Brien and Mike Woolverton, "World and U.S. Fossil Fuel Supplies," Agriculture Marketing Research Center Renewable Energy Newsletter, December 2009,
http://www.agmrc.org/renewable_energy/energy/world-and-u-s-fossil-fuel-supplies (Accessed 9-23-10).
284. US EIA, "Annual Energy Outlook 2013",
http://www.eia.gov/forecasts/aeo/electricity_generation.cfm (Accessed 1-17-14).

and the cost of *new* coal-fired power plants.

Social costs include the alleged impacts of human-caused global warming, premature deaths, and increased healthcare spending tied to *the dirty fuels of the past*. Unless we're willing to bias our thinking, however, we can't include these alleged costs without also including the social costs tied to "fuel poverty." We can't forget that CO_2 has always been a climate follower, indoor air quality is now a greater threat to our health than outdoor air quality, and South Koreans live much longer and enjoy a higher quality of life than their relatives in North Korea—despite having the most polluted cities of any developed nation. We can't ignore the social and health-related *benefits* derived from low-cost energy.

The other cost environmentalists understandably like to discuss is the cost of *new* coal plants. The key word is *new*. The DOE indicates that new coal plants produce energy at a cost of ten to twelve cents per kilowatt-hour ($0.10 to $0.12/kWh).[285] This cost is two or more times higher than the cost of energy from most *existing* coal-fired power plants and more than the cost of energy from *new* wind generation.

For wind, the key word is *energy*. We'll explain why in the following sections, but let's first explain why existing coal plants are far cheaper than any *new* source of electric energy production, including new renewable generation.

Most *existing* coal plants were financed long ago. We've already paid for their construction, so the only costs tied to energy from these plants are the cost of fuel, maintenance, and overheads. Installing any new generating facility carries a twenty to thirty-year premium to pay for the cost of construction. The cost of this premium is a cost the media and environmentalists have completely ignored. To understand the importance and the impact of this premium, we need to understand the difference between electric *capacity* and electric *energy*.

The Cost of *Capacity* versus *Energy*

Electric *energy* is the commodity we all know and understand. It's the real-time delivery of electricity that powers our lights,

---

285. US EIA, Ibid.

appliances, and electronics. Electric *capacity* is different. *Capacity* is provided by "steel-in-the-ground" generating assets that are capable of providing *energy* "on demand." In other words, *capacity* is provided by power plants that can reliably start, stop, and change loads whenever customers change their electric usage. Wind and solar technologies can't do that.

To understand the cost of *capacity*, consider what happens to your electric bill when the local electric company installs any new power plant. Your bill increases to pay for financing the cost of construction, right? Avoiding new construction is one of the best ways to maintain low electric prices. That's one reason why insulation, efficiency, and conservation are far superior investments than building new generating facilities.

More importantly, every region in the US has excess electric *capacity*. In other words, there's little need for large investments in *new* power plants.[286] As a result, when we mandate the installation of renewable technologies, we're mandating cost increases that aren't fulfilling an actual need. We're reducing our use of fossil fuels and that's terrific, but in the process we're significantly increasing our cost of electricity.

Capacity and the Family Car

Perhaps the best way to understand the cost of electric *capacity* is to think of it as though we're replacing the family car. In this case, our existing car (our existing electric infrastructure) is in good condition and it meets all our needs. It will take us anywhere we want to go, whenever we want to go there. The only reason we're looking for a new car is our desire for better gas mileage, a.k.a., our quest for cleaner, more sustainable energy. Because that's a terrific goal, we now find ourselves in the dealer showroom eying that shiny new car (wind turbines and solar panels).

Here's what the new car dealers (the media and the environmental lobby) haven't told us. Our existing car can run at top speed twenty-four hours a day, seven days per week, and 365 days

---

286. Reserve electric generating capacity helps keep the lights on, June 1, 2012; EIA, http://www.eia.gov/todayinenergy/detail.cfm?id=6510# (Accessed 1-19-13).

per year except for infrequent trips to the shop to keep it in top running condition. In other words, we can drive our old car roughly 90% of the time (existing fossil-based power plants can routinely produce 90% of their rated *capacity* over a year's time). Our new car choices (wind and solar energy) don't come close to matching this level of *energy* production.

Annual average *energy* production (miles driven) from wind is roughly 30% of its rated *capacity* and for solar it's roughly 20%.[287] New wind technologies can annually achieve over 40% of their rated *capacity*, but only when located in the windiest locations. Natural wind patterns, local topography, and the Endangered Species Act limit the availability of such locations. Similarly, solar panels can achieve more than 20% of their annual rated *capacity* in southern locations, but less than that in northern locations.

The bottom line is that we can only drive our new wind-car 30% to 40% of the time and our new solar-car 20% of the time. If we expect to drive the same number of miles we've driven in the past (if we want our new electric infrastructure to provide as much *energy* as our old one), we'll have to buy two to three new wind-cars or five new solar-cars. The cost impacts are tremendous, but they don't stop there.

The cost of renewable energy gets worse when we consider electric *reliability*. That's because no matter how many new cars we buy, we can't guarantee that our new cars will start when we need to drive somewhere. In other words, we can't trade in our old car. I've heard it argued that the sun is always shining and the wind is always blowing—somewhere. It's difficult to argue with such a bulletproof statement. At the same time, however, we have to realize that this statement doesn't equate to a reliable supply of electricity from wind and solar technologies.

Wind, for example, is most active at night and during the spring and fall: the exact times when our need for electricity (our need to drive) is at its lowest point. When our need for electricity is greatest,

---

287. US Wind Capacity Factor Analysis, January 4 2013, IHS Emerging Energy Research, http://www.emerging-energy.com/Content/Document-Details/Wind/US-Wind-Capacity-Factor-Analysis/1472.aspx (Accessed 1-11-13).

wind performs at its worst—if it performs at all. Consider the US heat wave of 2012. High pressure settled over multi-state regions in the Midwest and the winds remained relatively calm month after month. If we relied on blanketing the Midwest—the Saudi Arabia of wind—with wind turbines to meet our energy needs, we would've been out of luck. Solar panels would have performed at their best during this heat wave, but only during daylight hours and the national average production level would have remained around 20% of rated *capacity*.

Most of the studies promoting the superior economics of renewable energy fail to include the cost of buying two to five new cars—and still maintaining our old car. None of those studies consider the cost of electric *capacity*. They also fail to discuss the billions we're spending to build transmission lines to get wind energy from the Midwest to population centers in the East.

Too Much of a Good Thing

California law will soon require renewable resources to supply 33% of the state's electric *energy*. That means Californians will be paying to install two to five times more renewable *capacity* than this 33% *energy*-target. That's a huge capital investment. It may allow us to feel good about our decisions, but it will cost more than simply adding all those renewable resources. An article from the *Los Angeles Times* explains why.

The article states that California is being forced to install new fossil-based power plants to fill in the gaps when renewable generation isn't available. Industry insiders know that when large amounts of wind and solar energy are used in the same region, fossil plants must be ready to increase energy production instantaneously when clouds pass overhead or the wind slows. They must also reduce energy production instantaneously when winds increase or the sun breaks through the clouds. This instant response is required to maintain transmission system *reliability* and avoid power outages. As the amount of wind and solar generation increases in a region, the number of fossil-plants providing back-up energy must also increase.

To provide the necessary instant response, fossil-plants operate above their minimum level of production and below their optimum level of production. This allows them to quickly increase or decrease generation to match changing energy output from renewable resources. Like a car idling, fossil plants are inefficient when operating at low levels of production. Unlike cars, electric plants are most efficient when running near their maximum output (top speed). As a result, these back-up fossil-plants are burning more fuel, producing less energy, and emitting more pollution per unit of energy than they normally would. [288] It's an added cost and an environmental concern we seldom consider.

Energy Storage: The Missing Technology

Developing large-scale energy storage technologies would eliminate the need for fossil-plants to supply back-up energy for renewable resources. With energy storage, we could store excess wind energy produced during periods of low electricity usage and release it when it's needed. Of course, to replace our old car, we'd still need to buy several new cars plus we'd also need to buy a large number of new garages to store those cars. Replacing fossil fuels and our existing electric infrastructure won't be cheap.

To repeat an earlier question, are we certain we must find a way to eliminate our need for *costly* fossil fuels?

The First Law

We opened this chapter with a quote describing the First Law of Thermodynamics (the First Law). This law of nature tells us that energy cannot be created or destroyed. When we burn a gallon of gasoline, we don't destroy the energy; we transform it into work that moves our cars down the road. When we eat food, we transform the energy from calories into body heat, heartbeats, and simple muscle movement.

The First Law tells us that all the energy we'll ever be able to utilize already exists. Of course, we can't be certain the science surrounding the First Law is complete, but as far as we know today,

---

288. Ralph Vartabedain, "Rise in renewable energy will require more use of fossil fuels," *Los Angeles Times* (December 9, 2012), http://articles.latimes.com/2012/dec/09/local/la-me-unreliable-power-20121210 (Accessed 3-2-13).

there is no magic foo-foo-dust. All we can do is find new ways to transform existing energy into energy we can use. The First Law limits our options. It also tells us that any source of energy we utilize, even if it's renewable, will require us to take something from Mother Nature.

Unforeseen Consequences

We love the promise of renewable energy—as we should. However, today's renewable technologies are not without environmental sins. Today's large wind turbines rotate more slowly than the small, fast-spinning models that dominated early wind generators. It seems birds could easily avoid these slower-spinning blades, but the outer few feet of today's wind turbine blades are traveling at speeds from 180 to over 230 miles per hour.[289] For birds, today's wind turbines are like giant VegOMatics! An article in the *Smithsonian* magazine indicated that wind turbines kill 138,000 to 328,000 birds each year.[290] Over 80,000 of these deaths are birds of prey including bald and golden eagles—species that are protected under the Endangered Species Act.[291] According to *Save the Eagles International* and a variety of other organizations, wind turbines pose a far greater threat to endangered birds and highly beneficial species of bats than most believe.

Of course, birds die all the time by flying into windows and striking our windshields as we drive, so this is no big deal, right? Surely, the benefits of wind energy far outpace a few dead birds.

In yet another double standard for environmentalists, in 2013, after Duke Energy was fined $1 million for wind farms that killed 14

---

289. American Wind Energy Organization (AWEO), "Size specifications of common industrial wind turbines," http://www.aweo.org/windmodels.html (Accessed 1-11-14).
290. Rose Eveleth, "How Many Birds Do Wind Turbines Really Kill?," *Smithsonian* magazine, December 16, 2013, http://www.smithsonianmag.com/smart-news/how-many-birds-do-wind-turbines-really-kill-180948154/ (Accessed 1-10-14).
291. Robert Bryce, "We have to kill eagles with wind turbines in order to save them," *Wall Street Journal* (October 10, 2013), http://stream.wsj.com/story/latest-headlines/SS-2-63399/SS-2-351849/ (Accessed October 28, 2013).

golden eagles and 149 other protected birds,[292] the White House began allowing permits to be issued giving wind farms the right to kill Endangered birds for thirty years.[293] This action protects wind developers from prosecution and fines, removing a possible impediment to investments in wind power. We'd absolutely crucify the politicians granting such egregious protections for Big Oil, or any Big Business other than the *environmentally friendly* wind industry. Apparently, there's no end to the double standards we'll accept when it comes to the environmental lobby.

Beyond bird kills, the First Law introduces a much greater concern. Wind turbines literally knock the wind out of the wind. They slow natural wind currents close to Earth's surface. If we saturate the landscape with wind turbines, are we certain we won't disrupt weather patterns or potentially create an unforeseen climate disaster? That's not a hypothetical question.

NASA satellites indicate that wind farms increase Earth's surface temperatures in regions with a lot of wind generation. Atmospheric scientist Liming Zhou warned, "As wind farms become popular and much more widespread . . . they 'might have noticeable impacts on local-to-regional weather and climate.'"[294]

It's an issue few have mentioned, but it's a legitimate concern. Are today's renewable technologies really the long-term answer we've been seeking?

Energy Crisis #1: Beyond Coal

Since the early 1970s, environmentalists have been waging a war against coal. The Sierra Club called this war "Beyond Coal." They've won that war. Based on the specter of global warming, the

---

292. Pete Danko, "Wind Turbine Bird Deaths Cost Duke $1M," *Earthtechling.com*, http://www.earthtechling.com/2013/11/wind-power-bird-deaths-bring-1-million-levy/ (Accessed 1-11-14).
293. *PBS Newshour* (*Associated Press*, December 6, 2013) "Wind farms that kill bald eagles are now protected from prosecution," http://www.pbs.org/newshour/rundown/2013/12/wind-farms-that-kill-bald-eagles-are-now-protected-from-prosecution.html (Accessed 1-11-14).
294. Robert Lee Hotz, "Large Wind Farms Increase Near Ground Temperatures," *Wall Street Journal* (April 30, 2012).

EPA has announced new regulations that ensure we'll soon abandon our supply of abundant, cheap coal. As a first step in the war on coal, it's projected that by 2020 the US will be forced to retire 49 gigawatts (49 million kilowatts) of coal-fired electric *capacity*.[295] Using renewable resources to replace the *energy* provided by these retiring units will require electric ratepayers to foot the bill for installing 98 to 245 gigawatts of new renewable energy facilities. Depending on the blend of wind and solar energy used as replacements, the cost would range from $216 billion to over $1 trillion. In addition to that cost, we'd also have to spend more to install new non-coal, fossil-fueled generation to fill in the gaps when the wind isn't blowing and sun isn't shinning.

The cost impacts don't end with simply replacing the *energy* and *capacity* of retiring coal plants. In the process of abandoning coal, we'll not only drive up the cost of electricity, we'll also increase our use and the price of both natural gas and oil. Because natural gas and electricity are potential alternatives to oil, we'll ultimately make it more expensive to replace oil for transportation.

Our policies are creating the very energy crisis we hoped to avoid. Contrary to popular claims, moving to greener energy policies isn't going to stimulate job growth and create a vibrant economy. It will accomplish the exact opposite. It also does nothing to *sustain* our ability to access affordable energy.

Our World without Coal

The record setting cold wave of early 2014 provided a preview into the energy future we'll encounter thanks to the Sierra Club's war on coal. During this cold wave, many solar panels were covered with snow. The covered panels produced nothing. The blades on wind turbines were covered with ice forcing many to stop operating. Demand for natural gas for heating also skyrocketed and prices followed. In Texas, where natural gas serves as the major source of electricity, the cost of natural gas jumped to $90 per unit, some twenty times above normal. Wholesale electric energy prices spiked

295. US EIA, "Projected retirements of coal-fired power plants," July 31, 2012, http://www.eia.gov/todayinenergy/detail.cfm?id=7330 (Accessed 12-29-13).

to $5.00 per kWh—more than 100 times higher than the production cost of *energy* from most existing coal plants. Similar spikes in the price of natural gas and electricity were repeated in multiple regions with each Polar Vortex experienced in 2014.

Again, are we certain environmentalists are right when they say we must stop relying on *costly* coal?

Energy Crisis #2: Beyond Natural Gas

Remember when natural gas was the clean energy choice? Apparently, that's no longer the case. Having won their war against coal, the Sierra Club has a new war. It's a war they call "Beyond Natural Gas."

If history repeats itself, the Sierra Club will win their war on natural gas and we'll soon be walking away from it as well. We'll still have plenty of coal and natural gas; we just won't be allowed to use those resources. We'll be left with a decimated economy and the first real energy crisis we've ever known. We'll be left to rely on those two to five new cars that may or may not start when we need to drive somewhere. We'll all become very familiar with the term "fuel poverty."

Energy Crisis #3: Beyond Oil

In all my research, I haven't found the Sierra Club announcing a war called "Beyond Oil." Instead, they've announced their goal of ridding the world of fossil fuels by 2050. It's a goal we'd all love to achieve, but can we afford to get there? Will our quality of life improve, or will "fuel poverty" steal our futures, decrease our standards of living, and ultimately harm our health?

Fracking

In 2013, fracking was still a relatively new technology for recovering previously unrecoverable reserves of oil and natural gas. It was also rapidly becoming the prime target for environmentalists.

For example, the environmental lobby now claims that natural gas from fracking produces more global warming than burning coal.[296] This claim is based on EPA estimates of the volume of methane that escapes during fracking. New research, however, shows the EPA estimates are greatly exaggerated and fracking produces far fewer greenhouse gases than popularly claimed.[297] Nonetheless, the war against natural gas has begun.

There are numerous environmental concerns tied to fracking, but let's start by reviewing the benefits. Fracking is a breakthrough technology that is transforming the world of energy. The US has always had the world's largest supply of coal, but fracking greatly increased US reserves of oil and natural gas. As a result, the US now controls more economically recoverable fossil fuel resources than any other nation.

What does that mean for you and me? Fracking's impact on the cost of oil and natural gas can be spectacular. For example, prior to the advent of fracking, US natural gas reserves were in decline. Between 2003 and 2008, the price of natural gas increased from less than $4 per unit (per 1,000 cubic feet) to an average of $7.20 per unit. Prices spiked up to $15 per unit at one point and the long-term outlook called for a continuing rise in price. With fracking, economic supplies of natural gas have skyrocketed and prices have fallen back to below $4 per unit. In 2012 alone, the drop in price saved US households over $32 billion and saved all US natural gas users nearly $110 billion.[298]

It's difficult to overstate the significance of fracking. This is truly a world-changing technology. Most of us, however, don't view it in such glorified terms.

Fracking has more than its share of environmental concerns. It uses tremendous volumes of water, it can contaminate groundwater supplies, it's been linked to flaming tap water, and many studies

---

296. Ben Geman, "Study: Gas from 'fracking' worse than coal on climate," *The Hill* (April 10, 2011), http://thehill.com/blogs/e2-wire/155101-report-gas-from-fracking-worse-than-coal-on-climate (Accessed 9-1-13).

297. Russel Gold, "U.S. Overstates Leaks by Gas-Drillers, Says Study." *Wall Street Journal* (September 16, 2013), http://online.wsj.com/news/articles/SB10001424127887323981304579079 400039800412 (Accessed 10-1-13).

298. "Fracking and the Poor," *Wall Street Journal* Review and Outlook (September 7-8, 2013), A14.

indicate it may cause earthquakes. More recently, environmentalists have linked fracking to breast cancer. The war against natural gas is well underway and the environmental lobby is taking no prisoners.

The link between breast cancer and fracking has been compared to linking breast cancer to the carcinogens found in broccoli.[299] It's a ridiculous claim. Unfortunately, the other environmental concerns tied to fracking are very real.

You may have seen these concerns in Matt Damon's Hollywood film, *Promised Land*, or the Oscar nominated documentary by Josh Fox called *Gasland*. These films are powerful and influential. Scenes showing tap water catching fire are among the most influential images in the films. Flaming tap water is caused by the migration of methane gas into water wells. This is not a new phenomenon. Naturally occurring methane migration and flammable tap water have been documented as far back as 1936—long before fracking.[300] Still, fracking increases the likelihood of gas migration.

Fortunately, flaming tap water is easily rectified with proper water-well casings. It's not an insurmountable issue.[301] Industry is working on methods to reduce the potential for methane migration.

If we're worried about the vast amounts of water used for fracking, consider this: Watering our lawns uses twenty times more water than the fracking industry.[302] Are our priorities in the right place?

The fracking industry is developing technologies to reduce water usage and to combat the potential for groundwater

---

299. Kevin Begos (*Associated Press*, January 25, 2013), "'Fracknation' answer to 'Gasland'."

300. Julia Seymour, "Hollywood, hypocrisy and Matt Damon's anti-fracking film," (October 1, 2012), http://www.foxnews.com/opinion/2012/10/01/hollywood-hypocrisy-and-matt-damon-anti-fracking-film/ (Accessed 1-25-13).

301. Mike Soraghan, "Groundtruthing Academy Award Nominee 'Gasland'" (February 24, 2011), *New York Times* BusinessDay; Energy & Environment, http://www.nytimes.com/gwire/2011/02/24/24greenwire-groundtruthing-academy-award-nominee-gasland-33228.html?pagewanted=all (Accessed 2-20-13).

302. Rusty Todd, "Why The Grass Shouldn't Always Be Greener," *Wall Street Journal* (June 29, 2013), A15.

contamination. On-site evaporators have been developed to treat and clean spent fracking liquids. Industry is also developing eco-friendly alternatives for the fluids used in fracking. Ozonix[303] and SteriFrac[304] are two examples. These liquids use naturally occurring microbes to act as a disinfectant, replacing the use of chemical biocides. These new substitutes help eliminate the potential for groundwater contamination. These advances may also allow fracking fluids to be reused, which would reduce water usage and reduce the need for wells that are used to store the spent fluids.

On the other hand, how do we stop earthquakes? Fracking-related earthquakes have been linked to increased underground pressure created by the wells storing used fracking liquids. If industry can recycle those fluids, the need for storage wells will be reduced or eliminated.

What's the bottom line on fracking? A study at the Massachusetts Institute of Technology indicates the environmental concerns tied to fracking are "challenging but manageable."[305]

Make no mistake; the environmental lobby is determined to stop fracking no matter how safe it may ultimately become. Fracking is a remarkable, world-changing technology. Instead of fighting it, perhaps we should embrace efforts to find safe ways to use this technology. If not for ourselves, perhaps we should do it for future generations.

---

303. Ryan Fitzwater, "Ecosphere Technologies (OTC: ESPH): Offering a Cleaner Solution to Fracking," Investment U Research (December 20, 2011), http://www.investmentu.com/2011/December/ecosphere-technologies-otc-esph-cleaner-fracking.html (Accessed 2-20-13).

304. "100% Green Fracking Solution – SteriFrac," BusinessWire, January 10, 2012, http://www.businesswire.com/news/home/20120110005710/en/100-Green-Fracking-Solution---SteriFrac (Accessed 2-20-13).

305. Ernest J Moniz, Henry D Jacoby, Anthony J M Meggs, et al. "The Future of Natual Gas" (June 6, 2011), http://mitei.mit.edu/system/files/NaturalGas_ExecutiveSummary.pdf (Accessed 2-20-13).

Governmental Wisdom

Our reserves of fossil fuel are more promising today than perhaps any time over the last half-century. Unfortunately, our energy policies are largely based on old thinking. Here's what we knew when these policies were developed: We had the world's largest supply of cheap coal, we had an abundance of electric generating facilities, and our greatest energy concern was oil (transportation). Given this knowledge, what do you suppose our energy policies are designed to achieve? Our energy policies are hell-bent on abandoning coal and replacing our existing electric infrastructure. Meanwhile, we dabble with electric vehicles and mandate the use of small volumes of crop-based biofuels for transportation. As a popular beer commercial puts it: "Brilliant!"

More Government Wisdom

Our use of tax incentives to encourage desired behaviors often goes beyond the ridiculous. Tobacco is a prime example. We subsidize tobacco growers and then tax the hell out of their product because we don't want anyone to use it. It's social engineering on hallucinogens. Our approach to energy is similarly ridiculous. We subsidize both fossil fuels and renewable technologies.

We could throw numbers around all day and show two different results. Either our tax subsidies greatly favor renewable energy, or they greatly favor fossil fuels. As with any set of statistics, the different results are driven by the data included or excluded from the evaluation. For example, if we compare the annual tax write-offs for a company that invests one trillion dollars to a company that invests one billion dollars, the larger investments are essentially guaranteed to produce higher overall tax write-offs. Because the fossil fuel industry is far larger than the renewable energy industry, we should expect the total investment and annual tax write-offs for fossil-based energy companies to far exceed those of the renewable energy industry.

If we look at tax subsidies alone, however, renewable energy is the clear winner. In 2009, fossil fuels provided 78% of our energy, yet 77% of all energy-related tax subsidies went to renewable energy.[306] Today, wind projects receive direct subsidies of $23 for every megawatt-hour (1,000 kWh's) of electricity they produce during their first ten years of operation, plus additional incentives. By comparison, coal and natural gas receive roughly $0.64 per megawatt-hour.[307] To put that into perspective, we can look at an individual project to see the extent of the tax incentives we provide to encourage renewable energy.

Through direct tax credits, US taxpayers contribute either: 1) 30% toward the upfront cost to install a renewable energy project, or 2) the $23 per megawatt-hour described above. Many states also allow accelerated tax depreciation, which can save the project owners another 35% over the first five-years of operating a renewable project. Further, many investor-owned utilities provide rebates to help lower the cost of installing renewable energy. Finally, Renewable Energy Credits (RECs) are awarded for each unit of energy produced by a renewable resource. These RECs can be sold into cap-and-trade markets to further offset costs. You and I ultimately pay the cost of these subsidies and incentives.

Although it's a rare occurrence, solar project developers have told me these incentives can provide as much as 90% of the cost of their projects! My local electric supplier doesn't provide rebates and the value of REC's in my home state is low, so solar developers in my area only save around 65% of their project costs. With taxpayers contributing nearly two-thirds of the cost, however, none of these developers could offer solar energy pricing that was cost competitive with the cost of energy produced by my local electric company's existing coal plants. Again, we could argue statistics and subsidies all day long, but renewable energy isn't as economic as we're led to believe.

Although it's a legitimate argument to say that any benefits derived from tax subsidies for wind generation are shared by everyone, the same argument can't be made for subsidies and rebates for rooftop solar panels. Even with the subsidies, only those

---

306. Congressional Research Service, "Energy Production by Source and Energy Tax Incentives" (May 16, 2011).
307. Phil Gramm, "The Multiple Distortions of Wind Subsidies" (December 26, 2012), *Wall Street Journal*, A13.

with money (the rich) can afford to install solar panels and wait several years for a payback. Rooftop solar panels primarily benefit the rich. They also reduce the volume of energy sales for the local utility. As a result, utilities have to charge higher rate for each unit of energy sold in order to recover their overall cost: costs that are not significantly reduced by a reduction in energy sales. Rooftop solar panels increase income disparity and hurt the poor. Anytime the cost of energy increases, it disproportionally impacts the poor because the increased cost represents a higher percentage of their disposable income.

Better Returns?

Would our tax dollars provide better returns if we invested in research and development for the next generation of renewable technologies rather than funding the purchase of two to five new cars today? Would our tax subsidies be more valuable if we spent those subsidies installing LED lights to reduce our need to buy two to five new cars? Would our tax dollars do more to stimulate job growth, improve air quality, and spur the economy if we focused on transportation technologies rather than replacing our electric infrastructure? The answer to each question is "yes."

Electric Vehicles

Perhaps more than any alternative on the drawing board, electric vehicles (EVs) directly benefit two of our most pressing economic and environmental concerns: reliance on foreign oil and air quality in our largest cities. Unfortunately, today's energy policies will significantly increase the cost of electricity, making it all the more difficult to justify purchasing an electric vehicle. Our energy policies are grossly out of line with the fundamental facts.

With smarter policies, EVs could become a promising alternative for transportation, but they have their share of drawbacks. They have a short range of operation, recharging takes too long, replacement batteries aren't cheap, and no one has mentioned the environmental impacts of dealing with millions of old batteries. Despite these drawbacks, electric vehicles are an exciting option even if their limited range means they only replace oil for short commutes. Perhaps our tax dollars would yield more meaningful results if we spent them on battery technology and electric vehicles instead of replacing our existing electric infrastructure?

Natural Gas for Transportation

With minor modifications, internal combustion engines can burn natural gas rather than oil. That would be cleaner, and at current prices, it might also be cheaper than oil. However, today's low natural gas prices aren't likely to continue once we abandon coal and start using more natural gas for electric generation. How long will our supply of natural gas last and how much more will it cost if we use natural gas for home heating, electricity, *and* transportation? Natural gas is currently not the long-term solution we need for transportation. It simply trades one energy crisis for another. On the other hand, if we continue to use coal for electricity, we continue fracking, and we learn to tap frozen methane hydrates, perhaps we'd gain a few more centuries of affordable natural gas for both transportation and home heating. Unfortunately, the Sierra Club's "Beyond Natural Gas" war will eliminate this option from consideration.

Fuel Cells for Transportation

Instead of the internal combustion engine, we could use natural gas or hydrogen to power fuel cells for transportation. If natural gas isn't a long-term solution, hydrogen may be our best alternative for fuel cells.

The only emissions from hydrogen-powered fuel cells are heat and water. Additionally, we can manufacture hydrogen from water— a plentiful resource that's also renewable as long as it keeps raining and we don't deplete our oceans. It sounds like the perfect solution: cleaner air, no dependence on foreign oil, job creation to manufacture hydrogen here at home, and the flexibility and functionality of oil including quick refueling. Of course, building the necessary infrastructure won't be cheap. Hydrogen is also highly volatile, so safety is a significant concern.

If a hydrogen-powered future is our choice, the most likely path to manufacturing massive volumes of hydrogen is electrolysis: a process that requires a lot of electricity. Our current energy policies are significantly increasing the cost of this alternative. Additionally, it takes more energy to manufacture hydrogen from water than the energy provided by the hydrogen when it's used—a seemingly fatal flaw.

There may also be a few unforeseen consequences. All that water vapor might create flood zones downwind from major metropolitan areas, create breeding grounds for mosquito-borne diseases, and actually trigger catastrophic global warming. We might also create ice-covered roads during winter in northern regions. On the other hand, perhaps we'd generate plentiful clean drinking water and create beneficial rain forests that suck CO_2 out of the atmosphere—assuming that's a good thing and it won't throw us back into another ice age.

Biofuels

Biofuels promised a win-win. We could grow renewable fuel, save family farms, offset CO_2 emissions, and avoid an energy crisis. The win-win never developed.

Although the emissions are different than fossil fuels, burning biofuel pollutes and emits CO_2, just like burning any of those *dirty fuels of the past*. The promised CO_2 offsets are also dubious at best. The crops used to make biofuel act to absorb CO_2 while growing. Because we're continually replanting crops for more biofuel, we're told that atmospheric CO_2 won't increase. That may sound good in the popular media, but the land used to grow biofuels isn't barren land. It's covered with CO_2 absorbing plants whether or not we're specifically growing plants for energy. Additionally, it takes more energy to create biofuel than the energy we get by using the finished product—again, a seemingly fatal flaw.

The real fatal flaw is that without mandates for biofuel, much of the land currently used to grow fuel would be used to grow food. Mandating the use of biofuel increases the cost of food. We can't feed a starving world as it is, so growing crops for fuel instead of food seems not only irrational, but also inhumane.

Various attempts have been made to create biofuel from waste products and non-crop vegetation. These processes have proven more costly and less efficient than using crop-based plants. There's actually very little to like about biofuels other than their ability to reduce our need for imported oil. Even the environmental lobby has realized the problems tied to biofuels and their support for this alternative is rapidly vanishing.

Bacteria

Various forms of algae use photosynthesis to absorb CO_2 while growing into products we can use as fuel. President Obama has cited algae as the fuel of the future, but there are a few hurdles facing this technology. Algae only grow when the sun is shining. Burning the algae-derived fuel emits pollutants and CO_2. The surface area required to grow meaningful volumes of algae is staggering. In addition, the algae have to be protected from contamination: a potentially costly endeavor.

A process that absorbs CO_2 and generates oil substitutes sounds perfect, but more research is needed before this alternative can become a meaningful source of energy. Perhaps we should direct more tax subsidies away from buying two to five new cars and spend them on algae?

Another Alternative: The Plasma Arc

If you're looking for a sexy new electric energy technology, the plasma arc is it. This technology doesn't burn fuel; it vaporizes it. Fuel is loaded into a chamber that's essentially free of oxygen. Without oxygen there is no fire and emissions like furans, dioxins, sulfur dioxide, and nitrogen oxides are not a concern. Instead of fire, an electric arc creates tremendous heat that vaporizes the organic materials in the fuel, releasing gases that are then used to generate electricity. Unfortunately, burning those gases to produce electricity emits pollution and greenhouse gases, but technologies exist to remove most of these emissions.

This technology can utilize fossil fuels, but its real beauty is the ability to use any organic fuel source. This technology can literally eat landfills and generate electricity while producing virtually no pollution. The only waste byproduct is an impermeable glass that can be safely used in many industrial applications. There's a lot to like about the plasma arc.

Now the bad news: This technology isn't truly *commercialized*, and it is perhaps the most expensive source of electric generation available. It's so expensive the Energy Information Administration doesn't even list it as a commercially viable technology. That's why it's generally only used to destroy highly toxic wastes. The plasma arc is also highly inefficient. To generate the high temperature arc, this technology uses between 25% and 40% of the electricity it produces.

The good news is when we finally get tired of landfills, we have the technology to eliminate them—provided we can afford to do so.

Nuclear Energy

We can't, or at least we shouldn't, discuss clean air and sustainable energy without discussing nuclear energy. So here goes.

With Three Mile Island, Chernobyl, and the nuclear disaster in Japan, it's easy to see why we fear nuclear energy. New designs, however, eliminate the safety problems tied to Three Mile Island and Chernobyl. Some new designs also alleviate concerns with tsunamis and earthquakes. These earthquake-proof designs are small, self-contained nuclear plants that can be buried in the ground and abandoned in-place after expending their usable fuel. If we're worried about clean air and global warming, it's difficult to ignore the benefits of safe nuclear alternatives.

Instead of uranium, we can also use thorium to fuel nuclear power plants. Thorium is not considered suitable for nuclear weapons, and it's a plentiful resource that can extend the sustainability of nuclear energy for generations to come. Research and development efforts are underway to develop safe, thorium-based nuclear energy. Perhaps this research is a better place for our tax dollars than subsidizing the installation of today's renewable technologies and those two to five new cars?

Let's not forget nuclear fusion. Unlike nuclear fission, there are no radioactive by-products from fusion and fuel is virtually unlimited. Progress on this technology is slow and it's costly. If we can find a way to make fusion work safely and economically, it will be the sustainable, clean energy resource we've been seeking.

For numerous and understandable reasons, many of us oppose

nuclear energy. However, if we can perfect safe designs and recycle and protect stockpiles of spent fuel or develop a fusion alternative, we'll find a sustainable source of energy that doesn't pollute the air. If we want to be remembered as good ancestors, perfecting safe nuclear energy and a fusion alternative would be a great place to start. As much as we might fear nuclear energy, it's likely to remain a necessary evil for centuries to come.

Energy Solutions

It's not a sexy place to start, but our first priority should target those four items mentioned at the top of this chapter: insulation, passive solar, efficiency, and smart conservation. Investments in these areas provide the biggest energy and environmental bang-for-the-buck.

The next step may be uninspiring, but the solution for the next few centuries is to rely on every practical energy technology available, including coal, natural gas, oil, nuclear power, and renewable energy when and where each of these alternatives make sense. At the same time, we need a renewed focus on research and development for the next generation of energy technologies. At some point, the scale of this research may need to parallel that of The Manhattan Project: the project leading to development of the atomic bomb and atomic energy. Walking away from coal and natural gas will only accelerate the need for such urgent action. This might not be the solution favored by the media or environmentalists, but it should be the favorite choice for consumers, taxpayers, job seekers, and future generations.

Of course, the environmental lobby has a different solution. They prefer to pass regulations that limit access to fossil fuels and continually increase the cost of those fuels. By doing so, today's renewable alternatives become more cost competitive—but we'll still need to buy a lot of new cars.

The environmentalists' solution is death by a thousand paper cuts. Each new cut (each new regulated cost increase) is barely noticed, but over time, the mounting cuts (costs) bleed us into submission—and into "fuel poverty." This solution is well underway, and thanks to a biased media, inconvenient truths, and the training we've received since childhood, it's also the solution many of us favor today.

The Sierra Club makes no apologies. They proudly state their objective to achieve a total ban on fossil fuels by 2050. That's a wonderful goal, but do we really have the technologies needed to achieve that goal?

The environmental lobby tells us we do have the technologies to get there. They tell us we don't need the dirty fuels of the past because we can do it all with clean, renewable sources of energy. They tell us we can do it now and we can do it without destroying the economy, without losing jobs, without reducing standards of living, and without ultimately harming our health. Can we?

Chapter 21
We're from the Government;
We're here to Help

"We have to pass the bill so you can find out what's in it."
—Nancy Pelosi, US House of Representatives

No one needs to tell us that Washington is broken. The above quote simply proves the point. When Congress finally manages to pass legislation, lawmakers can't explain what it means, how it will work, or how much it's going to cost. It's a sad state of affairs.

On June 26, 2009, the US House of Representatives passed climate legislation titled *The American Clean Energy and Security Act of 2009.* It passed by a narrow margin, 219-212, but it passed the House all the same. The Senate refused to pass a similar bill so it never became law.

Bear in mind that the majority of the facts discussed in previous chapters were available to our House leaders in Washington. Our representatives knew—or should have known—that changes in atmospheric CO_2 have always followed changes in temperature; today's temperatures are below the average of the last 10,000 years; melting ice caps and rising oceans are nothing new; the UN along with many others have strong motives to convince the world that CO_2 is a dangerous pollutant; and our recorded history of temperatures began at the precise time the Little Ice Age ended.

House representatives had the responsibility to consider these things, yet they ignored fundamental facts and followed party lines to pass climate legislation. Whether you agree or disagree with the need to reduce CO_2 emissions, there's a more troubling story behind passage of this bill.

A Biased House

Prior to the final vote on this climate legislation, Al Gore was asked to address the committee charged with finalizing the bill's language. At the same time, some committee members requested that Lord Monckton of the UK also be allowed to address the committee. Lord Monckton was one of the expert witnesses who testified against the use of *An Inconvenient Truth* in UK classrooms. The committee refused to allow Lord Monckton to testify and heard only from Al Gore.[308] One reason given for objecting to Lord Monckton's testimony was the fact that he isn't a scientist. Of course, Al Gore is also not a scientist, so this objection was purely political. It was also irrational, but that's our Congress in action.

This House committee was biased from the start. They wanted no part of a legitimate debate. They weren't interested in the truth. They were interested in pursuing political agendas and they rushed to judgment in order to give the President something he could present at the upcoming 2009 UN Climate Conference.

Although many agree with the need for this legislation, it's difficult to imagine anyone supporting the committee's refusal to hear the whole story. That part is troubling enough, but what's more troubling is the economic cost associated with the legislation. If this climate bill had been signed into law, it would have become the most costly piece of legislation in US history.

The Cost of Climate Legislation

The climate bill passed by the House included a cap-and-trade program for greenhouse gases. As previously mentioned, that's equivalent to a hidden tax on practically everything we consume. The number of allowances available for this cap-and-trade market decreased overtime, so this hidden tax would escalate year after year. The legislation would have become the most economically destructive program ever envisioned by Congress. Of course, many House Representatives didn't see it that way.

They said the legislation would create jobs and wouldn't cost

---

308. Marc Morano, "Flashback April 2009: Democrats Refuse to Allow Skeptic to Testify Alongside Gore at Congressional Hearing," *Climate Depot* (April 23, 2009).

consumers one thin dime:

> 'There should be no cost to the consumer,' House Speaker Nancy Pelosi (D., Calif.) said Wednesday. She vowed the legislation would 'make good on that [pledge].'[309]

That's a pleasant thought, but most of us live in the real-world. We know that when the cost of energy goes up, the cost of doing business goes up, and that cost doesn't magically disappear. It's passed on to consumers. More to the point, if Congress can't tell what's in a bill until after it's passed and implemented, how can any member of Congress claim a bill won't cost consumers?

The Small Business Administration projected that complying with climate regulations would cost the economy $1.75 trillion per year, or about 12%-14% of overall GDP and nearly 50% of total annual Washington spending in 2009. The US Energy Information Administration projected that a 70% reduction in CO_2 emissions would increase the price of gasoline 77%, kill more than 3 million jobs, and reduce average annual household income by $4,000.[310] Exactly how did our governmental leaders conclude that these impacts would create jobs, stimulate the economy, and yield no cost to consumers? Perhaps more importantly, why didn't the mainstream media mention this side of the story?

A Revenue Neutral Carbon Tax

In 2009, the marketing gurus for the environmental lobby were loath to call climate legislation a tax. They knew that would never sell on Main Street. Today, things have changed. For the first time in my memory, voters in 2012 supported an increase in taxes at the Federal level—not just for the rich, but also for all wage earners

---

309. Greg Hitt and Stephen Power, "Democrats Weigh Break for Utilities in Climate Change Bill," *Wall Street Journal* (April 24, 2009), http://online.wsj.com/article/SB124050061773748291.html (Accessed 6-21-09).
310. Larry Bell, "The Alarming Cost Of Climate Change Hysteria," *Forbes* (8-23-11), http://www.forbes.com/sites/larrybell/2011/08/23/the-alarming-cost-of-climate-change-hysteria/2/ (accessed 11-5-12).

above the poverty level. That alone, was an epic sea change in voter sentiment. Today, it's no longer taboo to suggest a carbon tax—especially if it's promoted as a *revenue neutral* tax.

It sounds great on paper. Eliminate Federal income taxes; replace funding with a carbon tax; and no one will blink an eye when the cost of everything increases because we'll all have more money in our pockets. We'll clean the air, save the planet, and live happily ever after. It's an appealing approach, but can we trust the government to keep their end of this bargain? As Henry Ford advised, we might want to check with the Native Americans before answering.

Climate legislation has been packaged a number of different ways to make it sound more appealing, including "cap and cash back," "pollution reduction refund," and "revenue neutral tax." No matter how it's packaged, it's social engineering on steroids. I've heard this social engineering explained as follows. The answer is simple; tax fossil fuel, use those revenues to subsidize clean energy, and let the free market do the rest! Sadly, this is the way many view energy policy and free markets. There's nothing remotely *free market* in this line of thinking. It's picking winners and losers. In this case, it's picking winners that can't meet our energy needs—at least not without buying two to five new cars and keeping the old one around for times when the new cars won't start.

The increased cost of energy from a carbon tax would no doubt encourage better energy-related behaviors, but as previously noted, the cost impacts don't stop with energy. A tax on carbon increases the cost of essentially everything that is planted, harvested, mined, transported, processed, assembled, or manufactured. It increases the cost to heat and cool homes, cook dinner, watch TV, charge cell phones, drive to work, use a computer, or buy practically anything. It's an environmentalist's dream, but a consumer's nightmare. It will also hit low-income families the hardest—despite claims to the contrary and despite programs designed to assist those families with the increased cost.

We've said this before, but it bears repeating. The US economy is 70% dependent on consumer spending. What happens to that economy when the price of everything we consume goes up, and at the same time, we're paying more for the energy we use every day?

We're told nothing will happen. We're told that it'll all work out because consumers won't pay income taxes. Do you really believe that? Our governmental leaders can't even tell us what's in the legislation they pass, much less ensure that "it'll all work out."

Further, what happens to Federal funding as we begin transitioning away from fossil fuels? With less carbon to tax, they'll have to find something else to tax. Perhaps they'll suggest a return to taxing our income—assuming we have any income left to tax.

If you liked what the 1970s Arab Oil Embargo did for inflation, jobs, and the economy, you'll love what a revenue-neutral carbon tax will do for your standard of living.

Losing the Race

When our political leaders want to spur us to action, they talk about a race. We've seen the space race, the arms race, and now the clean energy race. Politicians tell us we're losing the clean energy race to China. They tell us US leadership is at stake if we don't win this race.

These are influential appeals because nobody likes to lose. When US politicians were in a full-court press to pass a climate bill in 2009, the *race* became a rallying cry to build support for the legislation. However, and as previously mentioned, in 2009 China was installing the equivalent of the entire fleet of US coal-fired power plants every three years and those plants weren't built with the pollution controls required in the US. Exactly how were our leaders determining the winner of this clean energy race?

To their credit, China is installing a lot of clean energy, but it's not because it's more economical or because they're trying to win any race. They're installing clean energy to pacify the rest of the world, and as we've already seen, because they're not always the ones paying for it. European countries have spent billions on clean energy projects in China in an effort to earn those CDMs tied to the now expired Kyoto Protocol.

Our government leaders are right in one respect: we are losing the clean energy *jobs-race*. China is currently the world's largest producer of solar panels and is leading the way in battery production for electric vehicles and energy storage.[311] The technologies to manufacture car batteries, energy storage devices, and solar panels are available to anyone who chooses to produce them. These components can be and are being produced in the US today, yet China is the leading manufacturer. They're the leading manufacturer because overly burdensome regulations make it more costly to manufacture these components in the US. How will climate legislation cure that problem? It will only serve to make manufacturing more expensive on our shores—and that's no way to win any race.

We're from the government. We're here to help.

---

311. Christian A. DeHamer, "The World's Biggest Polluter Looking for Green Profits," *Energy and Capital* (October 23, 2009), http://www.energyandcapital.com/articles/china-green-investment/980 (Accessed 3-11-10).

Chapter 22
What It All Means

"Our greatest responsibility is to be good ancestors."
—Jonas Salk

Perhaps no one has ever summarized the role of humanity more simply and more powerfully than Jonas Salk's quote above. It is our responsibility to protect the environment for future generations, but there's a point of diminishing returns—a point where our exuberance for environmental perfection slows economic activity and lowers standards of living. The resulting damage carries forward into the future, harming the opportunities and prospects available to our children and their children. Jonas Salk was a wise man. He was also a scientist, and he would have agreed with Dr. Patrick Moore: ". . . [environmental] stewardship requires that science, not political agendas, drive our public policy." Unfortunately, that's not how our policies are driven today.

Our beliefs and our policy choices are being driven by thirty-second sound bites: "The debate is over," "we're losing the race," and "we must eliminate the need for 'costly' coal." Clever marketing, politicians who think they have a better plan, and a biased media combine with the training we've received since childhood to deceive us all. We're not allowed to hear the whole story. Fundamental facts are hidden from our view. Science is corrupted in the name of achieving political agendas. These are the *facts* defining our understanding of energy, the environment, and their combined impacts on human health and the economy. This is not the path to informed decision-making, but it's the path we're forced to follow today.

When one study by one doctor convinces parents around the world to stop vaccinating their children against easily preventable deadly diseases, something is wrong with the news we're given.

When the world comes together to halt a program that saves millions of lives each year (DDT), something is wrong with the news we're given. When we support policies that create the very energy crisis we hoped to avoid, something is wrong with the news we're given.

We can't make wise decisions if we aren't given all the facts—and we can't be good ancestors if we can't make wise decisions.

The Environmental Lobby

We gave the environmental lobby our trust, our hearts, and our votes. In return, they gave us statistics, inconvenient truths, and intentional deception. They told us they had undeniable scientific proof that humans cause global warming. They told us the overwhelming majority of scientists agreed that humans are responsible for global warming. These are claims we can now reasonably challenge. They promised that green jobs would boost our economy—claims refuted by the real-world experiences of Europe and California. They warned that a small increase above current levels of CO_2 would cause unstoppable global warming—a claim that history shows cannot be true. Together with their marketing gurus, they've trained us from childhood. They've won their *Wag the Dog* marketing war using a philosophy that believes the facts "need not be spoken" and myth can be powerfully persuasive.

The environmental lobby has created artificial energy shortages where none existed. They've killed the future of our most abundant and lowest cost energy resource. They oppose large-scale farming and efficient livestock operations that provide economic nutrition for a starving world. They oppose solar energy in deserts, Golden Rice, and chlorinated drinking water. They oppose the things that have always improved human standards of living, increased longevity, and improved human health.

Historically, the term used to describe irrational environmentalism was *NIMBY* (Not in My Back Yard). That term evolved into *BANANA* (Build Absolutely Nothing Anywhere Near Anything). The term most descriptive of today's green movement is *NOPE* (Not On Planet Earth).

The environmental lobby has grown too powerful. Their

promotion of human-caused global warming is their biggest endeavor to date: It's the lie that goes too far. Together with the UN and the media, environmentalists have corrupted the very foundation of science. They called their deception an inconvenient truth. We should call it fraud.

The environmental lobby is no longer our friend or our benevolent protector. We can no longer trust organizations like the Sierra Club, Greenpeace, and a host of other seemingly well-intentioned groups—and that's a crime. These groups are filled with honest, concerned citizens who aren't environmental entrepreneurs and who aren't pursuing any political agenda other than protecting their health, the health of their children, and the health of the environment. They're fighting to oppose the power and influence of Big Money, they're fighting for a better world, and they're fighting based on the information they've been given: information that seldom tells the whole story.

Environmental organizations should be our watchdogs and our trusted guardians, but their leaders have turned into our enemy. It truly is a shame.

The environmental lobby makes no apology. Their ultimate goal is a world that operates more like North Korea than South Korea. It's a goal none of us should find appealing.

Kyoto and the UN

The Kyoto Protocol was a tool specifically designed to spur development in undeveloped nations at the expense of developed nations. It's the humanitarian thing to do, but it's based on preconceived ideologies and bad science. It does nothing of significance to stop global warming. We've been duped, and we bought it all.

Climate Legislation

Climate legislation is a hidden tax on practically everything we consume. It's a huge source of revenue for Big Government and an economic disaster in the waiting. If you liked what credit default swaps did to the world economy, you'll love what climate legislation will do to your disposable income. If you think Congress has wisely used our Social Security funds, you'll love what they'll do with a revenue-neutral carbon tax. If you trust our government to protect our best interests, remember the Native Americans.

The Media

The environmental lobby has broken our trust, but we should have seen that coming. We know what they stand for, so we can't blame them for doing exactly what they told us they would do. There are no similar excuses for the media. The media's coverage of global warming and practically every topic related to health, energy, and the environment has been grossly misleading.

Whether the media's misinformation is driven by incompetence, profit seeking, or an inherent bias, the media—the *news* media in particular—must be held to a higher standard of fairness and completeness. Investigative reporting is apparently a thing of the past. Instead of digging deep to find and explain the fundamental facts, it seems today's journalists only seek the sensational side of the story. When it comes to the media, we could rephrase a quote we saw in the first chapter: *There are three kinds of lies: lies, damned lies, and the media's version of how we should live our lives.*

The media's bias is unconscionable and it's an obscene abuse of their power and influence. The media deserves the totality of our wrath.

Sustainable Energy

We will run out of oil, natural gas, and coal. We'll eventually meet the environmental goal of a world free of fossil fuels. There's no reason to rush the arrival of that point in time. We've proven we can burn fossil fuels and still clean the air. With today's alternative technologies, we can't abandon fossil fuels and expect our economies to hum along without disruption. We have a few centuries' time to develop the technologies that will provide both an economic and a sustainable source of energy for future generations. Let's use that time wisely to further develop alternative technologies while maintaining strong economies, increasing standards of living, and improving our quality of life—not to mention improving our health and longevity in the process.

What it Means for Us

Affordable energy is far more important to our health than we've grown to believe. Energy is a fundamental price driver underlying the cost of practically everything we need, use, and consume. We have a choice regarding our cost of energy. We can continue to utilize the low-cost fossil fuels we control, and—as we've proven over the past forty years—we can use those fuels while continuing to clean the air, or we can follow the environmental path and walk away from centuries of low cost energy while cleaning the air even further. We've been trained from childhood to always pick cleaner air. Are we certain that's always the right choice?

Yes, we have a choice, but as previously mentioned, an organized few can control the disorganized many. The organized few have convinced us to hate fossil fuels. Should we? The organized few are costing us billions of dollars. Are the costs worth the rewards? The organized few have the media on their side and they have the ear of Congress. We can remain the disorganized many, or we can speak out. Will we? Will Congress hear from you? If not, the organized few will impose their will, you'll lose your chance to make a choice, and you'll be left to rely on those two to five new cars that may or may not start when you need to drive somewhere. Perhaps we should all send a copy of this book to our Congressional leaders?

Global Warming

The debate is indeed over, and Al Gore lost. The UN, Hollywood, the environmental lobby, and the politicians desperately wanting to tax CO_2 have all lost.

Whether we call it *global warming*, *climate change*, or simply *the weather*, human activities are not the driving cause. Human-caused global warming is a myth: It's the lie that went too far. The only catastrophe linked to climate change is the unfolding economic destruction tied to mandated restrictions of CO_2 emissions. We simply cannot allow the charade to continue. We must speak out. Will we? Will your Congressional Representatives hear from you? If not, you have no room to complain if you soon find your family living in fuel poverty.

Being Good Ancestors

Without affordable energy, economies suffer, standards of living decline, our health suffers, and our lives are shortened. More importantly, without a strong economy, we can't afford the luxury of environmentalism.

Who are the nations that are providing sustainable forests, clean drinking water, and sanitary sewers? Who are the nations that have reduced air and water pollution despite growing populations and increased energy usage? The developed nations of our planet are the ones that have the economic means to clean the water and the air. As an economy erodes, so does that economy's ability to care for the environment. A strong economy is not only vital to our health, it's also the only way we can afford to protect the environment.

If we want to be good ancestors, we must pursue rational environmentalism, not political agendas. Yes, it's vital that we protect the environment, but striving for environmental *perfection* is not the answer, at least not with today's alternative energy technologies. Our goal must be to protect not only the world's environment, but also the world's economies. We can't achieve both goals if we only listen to one side of the story.

On an annual basis today, South Koreans burn 34 billion more cubic feet of natural gas, 135 times more oil, and 100 million more tons of coal than North Koreans; yet they live thirteen years longer, grow much taller, and enjoy a higher quality of life than their relatives in the North.

Given the choice, where would you prefer your grandchildren live?

Above: The author enjoying his favorite pastime, camping and canoeing on the crystal clear, spring-fed streams in his home state of Missouri. In the late 1970s, Randall joined the environmental group *Friends of the River* to help protect the vital ecosystems provided by our fragile rivers. He understands the urgent battle we face to protect our planet, but he also understands the pitfalls associated with many of the battles we're currently waging.

About the Author

Randall L Hughes has spent over thirty years as an energy engineer focused on environmental compliance and long-term planning for our energy future. As an Energy Planner, he monitors and evaluates all aspect of our energy infrastructure in great detail. What are the pros, cons, and life-cycle costs of new and emerging energy technologies? How will current and future Congressional actions and environmental regulations impact each technology? Can increased efficiency, conservation, or other actions reduce our need for energy at a lower cost than building new resources? What is the real cost of pollution? How will national economic activity and the growth of undeveloped nations impact the cost and availability of labor, fuel, steel, copper, and a host of other critical resources? How will today's technology choices impact consumers and the environment now and in the future?

Due to the sheer volume of data involved, Energy Planners rely on computer simulation models to project the wide range of possible future outcomes. These Planners understand the benefits, limitations, and liabilities tied to computer modeling.

Randall has worked with the EPA, the Sierra Club, and Congress to find energy solutions that blend our need to protect nature with our need for affordable energy. He provided members of Congress with cost projections tied to proposed greenhouse gas regulations. He also worked with environmental groups to evaluate and implement alternatives for reducing carbon dioxide emissions.

Randall understands there are smart ways to protect nature and there are ways that actually cause more harm than good. Many of the things we do in the name of environmentalism yield little benefit at great cost. The cure is often worse than the disease.

Few public policy issues are more critical or more personal than those dealing with energy and its impact on the environment, our health, and our personal finances. When it comes to these critical issues, however, we haven't been told the whole story.

You can contact Randall at PopularDeceptions@Yahoo.com. You may also want to visit the website, PopularDeceptions.com.

www.ingramcontent.com/pod-product-compliance
Lightning Source LLC
Chambersburg PA
CBHW051633170526
45167CB00001B/167